Risky Agricultural Markets

Westview Replica Editions

The concept of Westview Replica Editions is a response to the continuing crisis in academic and informational publishing. Library budgets for books have been severely curtailed. Ever larger portions of general library budgets are being diverted from the purchase of books and used for data banks, computers, micromedia, and other methods of information retrieval. Interlibrary loan structures further reduce the edition sizes required to satisfy the needs of the scholarly community. Economic pressures on the university presses and the few private scholarly publishing companies have severely limited the capacity of the industry to properly serve the academic and research communities. As a result, many manuscripts dealing with important subjects, often representing the highest level of scholarship, are no longer economically viable publishing projects--or, if accepted for publication, are typically subject to lead times ranging from one to three years.

Westview Replica Editions are our practical solution to the problem. We accept a manuscript in camera-ready form, typed according to our specifications, and move it immediately into the production process. As always, the selection criteria include the importance of the subject, the work's contribution to scholarship, and its insight, originality of thought, and excellence of exposition. The responsibility for editing and proofreading lies with the author or sponsoring institution. We prepare chapter headings and display pages, file for copyright, and obtain Library of Congress Cataloging in Publication Data. A detailed manual contains simple instructions for preparing the final typescript, and our editorial staff is always available to answer questions.

The end result is a book printed on acid-free paper and bound in sturdy library-quality soft covers. We manufacture these books ourselves using equipment that does not require a lengthy make-ready process and that allows us to publish first editions of 300 to 600 copies and to reprint even smaller quantities as needed. Thus, we can produce Replica Editions quickly and can keep even very specialized books in print as long as there is a demand for them.

About the Book and Authors

Risky Agricultural Markets:
Price Forecasting and the Need for Intervention Policies
Pasquale Scandizzo, Peter Hazell, and Jock Anderson

This book shows how decisions made by individual farmers influence the efficiency of agricultural markets. Unless farmers properly take account of the correlation between prices and yields in forming their price forecasts, competitive markets will often be socially inefficient, leading to misallocation of resources. The authors demonstrate that a simple and practical price forecasting rule, based on expected per unit revenue, is generally adequate to ensure efficient market behavior.

Time-series data from various countries are used to test the hypothesis that market supply is influenced by the correlation of price and yield as well as by lagged market prices. The importance of market inefficiencies in risky situations is shown to depend on the variability of yields, the nature of farmers' price forecasting behavior, the degree of private risk aversion, and the elasticity of demand. The authors suggest and evaluate three basic policy approaches governments may take when confronted with very inefficient markets--establishing production quotas, improving market information services, and implementing price stabilization schemes. They conclude by discussing implications of the study for the specification of agricultural supply models and for the economic appraisal of risky investment projects.

Pasquale Scandizzo is economic advisor to the Secretary of Budget, Rome, and teaches at the University of Bologna. Peter Hazell is a research fellow with the International Food Policy Research Institute. Jock Anderson is professor of agricultural economics at the University of New England, Australia.

Risky Agricultural Markets

Price Forecasting and the
Need for Intervention Policies

Pasquale Scandizzo,
Peter Hazell,
and Jock Anderson

Routledge
Taylor & Francis Group
LONDON AND NEW YORK

First published 1984 by Westview Press

Published 2019 by Routledge
52 Vanderbilt Avenue, New York, NY 10017
2 Park Square, Milton Park, Abingdon, Oxon OX14 4RN

Routledge is an imprint of the Taylor & Francis Group, an informa business

Library of Congress Cataloging-in-Publication Data
Scandizzo, Pasquale L.
    Risky agricultural markets.
    (A Westview replica edition)
    Bibliography: p.
    1. Agricultural prices--Developing countries--
Forecasting.   2. Farm risks--Developing countries.
3. Agriculture and state--Developing countries.
I. Hazell, P. B. R.   II. Anderson, Jock R., 1941-
III. Title.
HD1417.S23   1984   338.1'3'091724   84-19659

ISBN 13: 978-0-367-28615-6 (hbk)

ISBN 13: 978-0-367-30161-3 (pbk)

Contents

Tables and Figures

x

FIGURES

Preface

Our experience in getting this work into print epitomizes our view of the omnipresence of risk in economic environments, whether they be faced by farmers in the developing world or by authors in the industrialized world.

We began work on the topic during the early 1970s when two of us, Hazell and Scandizzo, were engaged in developing sectoral planning models in the Development Research Center of the World Bank. Our concept of working the ideas into a monograph took shape during 1977 when Anderson spent some time at the Center. However, our efforts in this direction since then have been rather piecemeal, reflecting emerging and usually conflicting interests, although we assembled much of the present portfolio of material in 1979 and early 1980.

Since then, others have been busy in related research activities, and our final thrust, which was really more of a parry, was in 1984, when we attempted to integrate, or at least recognize, the importance of these activities as they bear on our work.

The long-standing collaboration between Hazell and Scandizzo in researching risky issues was one of equals and the order of names on previously published works merely alphabetic. To correct any unintended bias implicit in the earlier convention, for this book we have reversed the order.

As with all products resulting from a sustained period of research, we have benefited from the advice and help of many. In the World Bank, John Duloy, Roger Norton (now of the University of New Mexico) and Graham Donaldson were instrumental in arranging financial and administrative support. Roger Norton was also a source of encouragement and a valuable and perceptive critic. We are also grateful to the International Food Policy Research Institute (IFPRI) whose Director, John Mellor, generously provided for Hazell's continued involvement in the research and writing of this book after he left

the World Bank in 1979, and for the preparation of the
final manuscript. We hasten to add that the views and
interpretations in this book are those of the authors,
and should not be attributed to the World Bank or to
IFPRI.

We have also benefited from the advice and comments
provided by many colleagues. Notable among these were
Wilfred Candler, Gerald O'Mara, Gershon Feder, Raymond
Byron, Carlos Pomareda, Ammar Siamwalla, Giancarlo
Moschini, Stanley Fisher, Colin Gellatley and William
Griffiths. We are also indebted to Richard Just who
contributed to the research underlying Chapter 3.

In the World Bank, Cynthia Hwa, Malathi Parthasarathy, and Vinh Le-Si provided expert research
assistance, and Helen Claverie and Leela Thampy typed
various earlier versions of the manuscript. Last, but
by no means least, we are indebted to Melanie Snyder and
Ding Dizon who gave generously of their time and skills
in preparing the final manuscript at IFPRI.

Some parts of this book initially appeared elsewhere
in somewhat different forms. Chapter 4 is based on
material originally published by Peter Hazell and
Pasquale Scandizzo in "Competitive Demand Structures
Under Risk in Agricultural Linear Programming Models,"
American Journal of Agricultural Economics, Vol. 56, No.
2 (May 1974), and in "Farmers' Expectations, Risk Aversion and Market Equilibrium Under Risk," American Journal
of Agricultural Economics, Vol. 59, No. 1 (February
1977). Some of this material also appeared in revised
form in "Risk in Market Equilibrium Models for Agriculture" by Peter Hazell and Pasquale Scandizzo and in
"The Importance of Risk in Agricultural Planning Models"
by Peter Hazell, Roger Norton, Malathi Parthasarathy and
Carlos Pomareda, both of which were published in The
Book of CHAC, Programming Studies for Mexican Agriculture, Roger Norton and Leopold Solis M. (editors),
Baltimore, Johns Hopkins University Press (1983). Chapter 5 draws on "Producers' Price Expectations and the
Size of the Welfare Gains from Price Stabilization," by
Pasquale Scandizzo, Peter Hazell and Jock Anderson,
Review of Marketing and Agricultural Economics, Vol. 51,
No. 2 (August 1983) and on "Evaluating Price Stabili-
zation Schemes With Mathematical Programming," by Peter
Hazell and Carlos Pomareda, American Journal of Agricultural Economics, Vol. 63, No. 3 (August 1981). Chapter
6 is a revision of "Project Evaluation in Risky Markets"
by Pasquale Scandizzo in Operations Research in Agri-
culture and Water Resources, Dan Yaron and Charles
Tapiero (editors), Amsterdam, North-Holland Publishing
Company (1980), and copyrighted by the International
Federation of Operational Research Societies. Permis-
sion to include material from all these sources is
gratefully acknowledged.

1
Introduction

Agricultural planning, whether at the sector, farm or project level, is invariably characterized by considerable risks. Risks may arise from many sources including unpredictable changes in consumers' tastes and in planners' policies but emerge primarily from the variability inherent in natural climatic and biological systems. Risk is a particularly pervasive phenomenon in the agriculture of developing countries and is deserving of careful attention in both policy formation and in the planning and appraisal of projects addressed to improving the lot of people in these countries.

IMPACTS OF RISK IN AGRICULTURE

While there is no controversy over the pervasive effects of risk in farming, only scant attention has been concentrated on dealing explicitly with these effects. The ploy of most agriculturalists, economists and others has been at worst to assume away such problems or at best to choose values of key parameters that are discounted in some informal way for any risks perceived.

Where there has been some formal attention given to risk, the focus has usually been at one of two rather disparate levels. First, the direct impacts on farmers, especially small farmers, have been emphasized. The causation has been insightfully decomposed into two components (a) the farmers' perceptions of risks and (b) their innate aversion to risk. Taken together, these components of farmers' decision making result in behavior that has variously been described as 'irrational,' 'conservative,' 'cautious,' 'safety-seeking,' 'rational,' or 'utility-maximizing,' to mention just a few. Whatever may be the observer's perspective, it is evident that analysts need to be aware of the prospects of such behavior and of the consequences for adoption and

1

performance in projects and of farmers' response to policy changes. Work at this level has been reviewed by Anderson, Dillon, and Hardaker (1977) and by several authors in Roumasset, Boussard, and Singh (1979).

At the second level, there has been much discussion of commodity stabilization schemes and here the affected parties of primary attention have been consumers and producers of foodgrains, and producers of exported industrial crops. Seemingly much of the motivation behind stabilization schemes stems from the notion that the fluctuations observed in prices and incomes are deleterious to the economic health of regions, countries and, indeed, the whole world. Economic theoreticians have developed a diversity of analytical models to provide rationalizations of the beneficial effects of enhanced stability. Work at this level has been reviewed by Turnovsky (1978) and related empirical work also summarized in Adams and Klein (1978). The work of Newbery and Stiglitz (1979, 1982), and especially the seminal contribution compiled in 1981, represents the most significant recent advance in the field.

Generally speaking, with these recent exceptions, there has been relatively little effort expended to study the economic linkages between individual producers confronted by risky production and prices and the markets in which these risky prices are formed. For instance, most analysts of the social efficiency of agricultural markets have failed to recognize that producers of most products must necessarily make production decisions and resource commitments without precise knowledge of the prices that will eventually be realized. At best they will have knowledge of a probability distribution of prices--in short, they are making decisions under risk.

OUTLINE OF THE STUDY

Of the few cogent studies of the type alluded to, two are by two of the present authors (Hazell and Scandizzo 1975, 1977)[1]. It is the first intention in this monograph to generalize these earlier results concerning the nature of producers' anticipations about market signals in risky markets and the consequent effects on social efficiency (economic welfare). Generalization is taken up in Chapter 2. The principle finding is that, in simple but plausible specifications of markets characterized by risky production and thus prices, producers need to take account of market information beyond that contained merely in the prices they have experienced if the market is to function optimally from a social point of view. In some simple, intuitively reasonable, characterizations of market structures,

it is a measure of unit revenue (involving simultaneously price and yield) about which producers should
formulate their market anticipations for such social
optimality to prevail. That is, the argument of the
producers' supply function is something different from
simply a price and should embody at least joint price
and yield combined influences.

Agricultural supply functions have been estimated
for numerous regions and products ever since agricultural economists embraced the emerging methods of econometrics over the past few decades. Typically, if not
almost invariably, the supply functions have been specified in terms of anticipated prices. In the light of
the theoretical results of Chapter 2, it is thus of
relevance to inquire (in Chapter 3) into the potential
empirical superiority of specifications predicated in
terms of anticipated revenues. Unfortunately, our data
in this review are only for past time series of observed
prices and quantities and it was not possible to deal
with producers' subjective anticipations as such. Perhaps it is thus not too surprising to establish in this
way only limited empirical support for a condemnation of
mere price anticipations in aggregate supply functions.

Undaunted by these results from the historical
world, we embark in subsequent chapters to explore some
implications for various aspects of agricultural policy.
Mathematical programming models of resource allocation
in agricultural sectors have been utilized by national
planners in a number of countries. Because they are
invariably models of producers' aggregate behavior in
risky markets, it is important that farmers' anticipatory processes are appropriately represented if planners
are to identify social optima. In this context, we are
able in Chapter 4 to demonstrate more directly the welfare consequences of alternative specifications of anticipations as well as to elaborate on apposite method-
ological procedures for dealing with revenue anticipations and farmers' aversion to risk.

In Chapter 5 we take up questions of stabilization
from both theoretical and numerical standpoints. We do
this from the perspective of representing producers as
confronting risk rather than merely instability and from
the particular vantage point of the theoretical results
of Chapter 2 concerning anticipations. ` Our findings in
this chapter underline the importance of better understanding of the anticipatory behavior underlying market
structures and give important clues as to the conditions
which must prevail for significant benefits to be in
prospect from attempts to stabilize markets.

Finally, in Chapter 6 we explore implications of our
findings in earlier chapters for shadow pricing in project appraisal. Again, the joint roles of prices and
yields in risky markets are given prominence. If shadow

prices are based purely on observed and projected prices, they will tend to be higher than is appropriate with the consequences of potentially distorting comparisons and selections of projects. We conclude that what is called for is judicious use of social cost benefit analysis and pragmatic methods to approximate project-specific revenues.

LIMITATIONS OF THE STUDY

Throughout our work we have employed the idea that the simplest representation that captures the essence of reality is best for our purpose. This has led us into using families of models that many readers will surely reject as too naive or too simplistic. Simplicity and relevance are intrinsically judgemental concepts and we stand by our judgement in presenting the results that we have.

Apart from the structure of our market models, the major simplifying assumption we have made is that social welfare can be approximated by summing measures of expected producers' and consumers' surplus. We are aware of the several limitations concerning this procedure but we resort to it on the grounds of (a) workability, (b) economy of assumptions, and (c) established tradition.

Much scope exists for kindred research of both theoretical and empirical orientation, as evidenced by the pioneering work of Newbery and Stiglitz (1981). We would like to see further work on general equilibrium models of multicommodity markets. So little empirical work of direct relevance to the issues raised here has been done that the field is largely untapped. We draw the line on our empirical work here because of the resource constraints that we faced. We look forward to the greater resolution of the uncertainties about producers' anticipations in their planning through applications of new methods and data that we have had neither the wit nor the knowledge to behold.

NOTES

1. Other works in which a similar stance is taken include those of: Muth (1961), Turnovsky (1973), Newbery (1976), Aoki (1976), Newbery and Stiglitz (1981), and Wright (1979).

2
Expectations, Welfare and Market Equilibrium

Hazell and Scandizzo (1975) have argued that competitive markets are socially inefficient when (a) production is risky and this risk enters the supply function multiplicatively, and (b) supply decisions are based on anticipated (that is, uncertain at decision time) prices, where anticipations about prices are formed independently of those about yield.

Subsequently, in response to Newbery (1976), Hazell and Scandizzo (1977) demonstrated under an analogous set of assumptions that, when decisions are based on anticipated (unit) revenues (that is, accounting for the joint distribution of prices and yields), the social inefficiency and its associated optimal distortion of the market both vanish. The inefficiency and distortion also vanish in the event that supply risk is additive rather than multiplicative (that is, intercepts rather than slopes of supply schedules are stochastic).

We deduce from these results that there is a need for empirical research into the nature of both production risks and producers' formulation of expectations before conclusions about market efficiency and optimal intervention policies can be drawn with confidence and relevance. The conclusions will have relevance for policymakers' decisions. Hazell and Scandizzo in their preliminary results also raised some theoretical questions about the generality of their findings in the context of possible extensions beyond the domain of simple linear and partial equilibrium models of a single market. These questions are also taken up in this chapter.

Our work towards a more general theory of market efficiency under risk is described here. First, a partial equilibrium analysis of a single market is considered, but without introducing any specific functional forms. This framework is used to provide a direct generalization of Hazell and Scandizzo's results, as well as to explore market efficiency in long-run equilibria.

Two complications are then introduced. First, the effect of having two groups of producers with less than perfect correlations between yields and second, the effect of joint production of risky crops. In both cases, interest is focused on whether the existence of correlations between groups of products leads to different types of rational expectations for market efficiency. Then a general equilibrium analysis of the problem is undertaken to examine whether a more complete treatment of income and wage effects would lead to changes in the rational expectation for a risky market.

PRINCIPAL RESULTS

The main finding is that, if producers formulate expectations about unit revenue (price times yield divided by mean yield), this is sufficient to guarantee market efficiency, provided also that the marginal utility of money is assumed to be constant. In this case, 'revenue expectations' have the properties (a) of being self-fulfilling expectations on average at market equilibrium, (b) of being consistent with long-run market equilibrium, and (c) of containing all the information necessary for maximizing the assumed concept of social welfare.

When the marginal utility of money is assumed to be other than constant, revenue expectations represent the best that decentralized decisionmakers can do when provided with market and production information. However, they do not lead to the assumed social optimum. In this more complex situation, there may be a case for policies involving direct intervention by governments.

Revenue expectations differ importantly from 'price expectations' (which typically are assumed in economic analyses) in that they lead to less production on average of products that are produced under conditions of risk. In later chapters, implications of revenue expectations are explored for both agricultural policy in risky markets, and some particular economic analyses such as supply analysis and shadow pricing in project appraisal.

PARTIAL EQUILIBRIUM ANALYSIS OF A SINGLE MARKET

Short-Run Analysis

Consider a market in which the quantity of a product demanded by consumers during a given period t, D_t, is related to the price of the product, P_t, in a deterministic way as

(2.1) $D_t = f(P_t)$.

Our exclusion of stochastic elements from the demand relationship is for reasons of empirical reasonableness and analytical simplicity. However we believe our subsequent results are probably robust with respect to changes in this assumption.

Producers in this single-product market confront a risky production function where, for simplicity, risk is assumed to have its source only in crop yield, (unknown at the time of resource commitment) with realized values denoted by ε_t. The (realized) production function (see also Pope and Just 1977, and Quiggin and Anderson 1979) can be written as

(2.2) $Q_t = h(Q_t^*, \varepsilon_t)$,

where the star in Q_t^* denotes that this is the quantity which producers anticipate marketing on the basis of their personal situations of available productive resources, family consumption demands, production costs including those associated with aversion towards risk, and perceptions of risk. In this simple model with an implicit planning horizon of one period, it is assumed that producers' perceptions consist of knowing enough about the distribution of yields to anticipate the mean, $\varepsilon_t^* = E[\varepsilon]$, and of anticipating information about returns in terms of a random variable P_t^* which might be thought of as a certainty equivalent price for planning purposes. Thus Q_t^* can be assumed to be a nonstochastic function of anticipated price P_t^* which is consistent with producers making decisions about resource use in order to maximize expected utility. From the theory of the firm under risk as developed, for example, by Magnússon (1969) and Sandmo (1971), producers could be interpreted as equating their certain marginal costs (including marginal costs of risk and risk aversion) to expected marginal revenue represented by P^*. In fact, as discussed below after presenting Equation (2.11), the strict separation of these environmental and preferential effects associated with risk in the cost functions is only straightforward when utility functions are of very particular types.

This leads to an alternative representation of the aggregate production function as an aggregate supply function in which producers' marginal costs are aggregated. This function can be written as

(2.3) $S_t = \varepsilon_t A_t = \varepsilon_t g(P_t^*)$,

where A denotes area allocated to production of the product in question and $A_t = g(P_t^*)$ is the deterministic component of supply, while S_t is the quantity actually supplied at anticipated price P_t^*.

The model introduced thus far can be closed in several ways. The simplest is to ignore storage and to introduce a market clearing condition,

(2.4) $D_t = S_t$,

and to complement this condition with reasonable assumptions that will ensure that the market behaves in a realistic way. One such constellation of assumptions is to fix the mean and variance of ε_t at $E[\varepsilon_t] = \mu$ and $V[\varepsilon_t] = \sigma^2$, to assume that yields are serially independent, which implies that $cov[\varepsilon_t, \varepsilon_{t-1}] = 0$, to attribute to the supply and demand functions the conventional shapes, $f'(P) < 0$ and $g'(P^*) > 0$ and, somewhat more restrictively, to assume that $cov[P_t^*, \varepsilon_t] = 0$. This last assumption is consistent with the idea that producers, in forming their anticipations about price, really do not have any particular information on the eventual realization of yield in each planning period. Note however, that no assumption is introduced about the independence of realized yields and prices. Indeed, these will be negatively correlated with downward sloping demand curves.[1] With yield evaluated at its mean, the supply function (2.3) can be interpreted as an anticipated supply function and written as

(2.5) $S_t^* = \mu A_t = \mu g(P_t^*)$.

Naturally, since yield is random or stochastic, actual price is stochastic. Its probability distribution depends on the distribution of both yield and anticipated price as these are revealed over time. The operation of the market is sketched in Figure 2.1 by depicting one arbitrary anticipation and realization. For simplicity, we will merely presume convergence in mean and variance of price in our markets--that is, that with the passing of enough time, the mean and variance of price settle down to fixed values (that is, $\lim_t E[P_t]$ and $\lim_t V[P_t] = $ constants as $t \to \infty$). The mathematical requirements for such convergence depend on the nature of f( ) and g( ) and are explored for the linear case in Bergendorff, Hazell, and Scandizzo (1974).

Note in Figure 2.1 that the multiplicative random yield rotates the supply function in a random fashion. At time t producers are assumed to have anticipated price P_t^* , and to plan so that in aggregate they produce an anticipated supply S_t^*. At that level of anticipated output, production costs are equal to the area cdfe, which is the area under the anticipated supply function between 0 and S_t^*. This area is also equal to area cdhe less area dhf, that is, the area

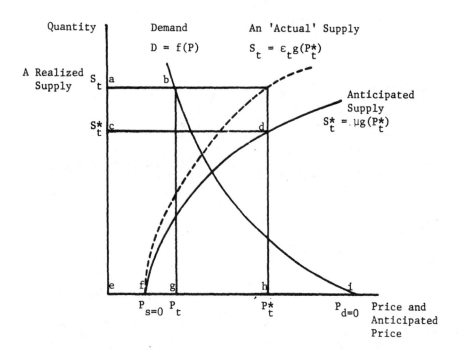

FIGURE 2.1
A Schematic Representation of the Assumed Market

$$P_t^* S_t^* - \int_{P_{s=0}}^{P_t^*} \mu g(P) dP.$$

However, when yield is realized, actual quantity supplied is $S_t = \varepsilon_t g(P_t^*)$, and the corresponding market clearing price is P_t. The realized producers' surplus, or ex post profit, is

$$(2.6) \qquad W_p = P_t S_t - [P_t^* S_t^* - \mu \int_{P_{s=0}}^{P_t^*} g(P) dP],$$

that is, actual revenue minus actual costs or, in terms of Figure 2.1, area abge less area cdfe. This is a more relevant measure of producers' welfare than is the ex ante surplus,

$$\mu\int_{P_{s}=0}^{P_t^*} g(P)dP,$$

typically used in earlier studies, because it captures
the windfall gains and losses in income, $P_tS_t - P_t^*S_t^*$,
that arises from forecast errors.

In the absence of knowledge of the complete map of
the tastes and preferences of consumers, their welfare
can be measured reasonably well by realized consumers'
surplus (Willig 1976). This is the area under the
demand function and above the realized price, namely

$$(2.7) \qquad W_C = \int_{P_t}^{P_d=0} f(P)dP,$$

or, in terms of Figure 2.1, area big.
Total welfare can then be approximated by summing
the producers' and consumers' surplus, that is, W =
$W_p + W_c$. Our selected criterion for measuring social
welfare in the equilibrium state of risky markets is
expected total surplus. Apart from our refinement of
measuring this in an ex post sense for the production
side of the market, there is a firmly established tradition of using surplus in efficiency appraisals of both
theoretical and empirical orientation (see for example,
Waugh 1944, Oi 1961, Massell 1969, Turnovsky 1974, 1976,
and Just 1978). From Equations (2.6) and (2.7), our
welfare measure is thus

$$(2.8) \qquad E(W) = E(W_t)$$
$$= E[P_t\varepsilon_t \ g(P_t^*)] - \mu E[P_t^* \ g(P_t^*)]$$
$$+ \mu E[\int_{P_{s}=0}^{P_t^*} g(P)dp] + E[\int_{P_t}^{P_d=0} f(P)dP].$$

The key issues raised in the Hazell and Scandizzo
analyses depend on the way in which P^* is formed. If
farmers form anticipations about price independently of
their anticipations about yield, the market may converge
to an equilibrium in which the anticipated price is, on
average, a self-fulfilling expectation. However, this
equilibrium may be inefficient in terms of maximizing
social welfare. Indeed, this was the finding in the

specific parameterization employed in Hazell and Scandizzo (1975). The question thus arises as to what is the optimal anticipated price which is not only self-fulfilling, but also rational in the sense of maximizing social welfare. Hazell and Scandizzo (1977) claimed that the optimal P^* is a revenue expectation, although this result was derived in a restrictive analysis that featured a linear model with multiplicative (or heteroscedastic) risk. An attempt is now made to generalize this finding using the less specialized market structure defined above.

The feature of present uncertainty and particular interest in this market is the nature of producers' anticipation about returns. This nature can be approached in two ways.

First, we can inquire as to the consequences of competitive producers following the traditional advice that they should maximize their returns by equating (expected) marginal returns and marginal costs. From the assumptions underlying the formulation of the supply function, marginal cost is assumed to have been equated with, and can thus be represented by P_t^*. Marginal revenue can be derived from total revenue, $P_t S_t$, by differentiating with respect to planned total output or quantity supplied, S_t^*. This evaluates as

$$(2.9) \qquad \partial(P_t S_t)/\partial S_t^* = P_t \partial S_t/\partial S_t^*,$$

because in the competitive environment assumed for producers, P_t does not vary directly with S_t^* (that is, $\partial P_t/\partial S_t^* = \partial P_t/\partial Q_t^* = 0$). Expected marginal revenue is thus determined as

$$(2.10a) \qquad E[\partial(P_t S_t)/\partial S_t^*] = E[P_t]E[\partial S_t/\partial S_t^*]$$
$$+ \operatorname{cov}[P_t, \partial S_t/\partial S_t^*],$$

using the standard result for the expectation of the product of two random variables. The term $\partial S_t/\partial S_t^*$ (or $\partial Q_t/\partial Q_t^*$) can be expanded from the assumptions stated in Equations (2.3) and (2.5) as $\partial(\varepsilon_t g(P_t^*))/\partial(\mu g(P_t^*)) = (\varepsilon_t/\mu)(\partial g(P^*)/\partial g(P^*)) = \varepsilon_t/\mu$ which is a dimensionless or standardized measure of yield with an expectation of unity, since $E[\varepsilon_t/\mu] = E[\varepsilon_t]/\mu = \mu/\mu = 1$. Thus (2.10a) can be simplified as

$$(2.10b) \qquad E[\partial(P_t S_t)/\partial S_t^*] = E[P_t] + \operatorname{cov}[P_t, \varepsilon_t/\mu]$$
$$= E[P_t \varepsilon_t]/\mu.$$

The final equality in (2.10b) follows directly from the expectation of a product of random variables and can be interpreted as an expectation of a unit revenue, $P_t\varepsilon_t = R_t$, standardized by mean yield so that the measure is in units of price. This first approach to exploring the nature of producers' anticipation about returns can now be summarized by the equation of marginal cost and marginal return in

$$(2.11) \qquad P_t^* = E[P_t] + cov[P_t, \varepsilon_t/\mu]$$

$$= E[P_t\varepsilon_t]/\mu$$

$$= E[R_t]/\mu.$$

Producers then should thus anticipate as their marginal return and cost not simply the average price, but rather expected unit revenue in which the joint effects of varying yield and price are captured. In our assumed market with a downward sloping demand curve, yield and realized price are negatively associated so that the co-variance term in (2.11) will be negative and therefore $P_t^* < E[P_t]$. This means that anticipated price should be set conservatively in the sense of being less than the expected price.

The result in (2.11) is a consequence of the particular 'multiplicative risk' specification of the supply function. If this is specified in an 'additive risk' but implausible way such as $S_t = g(P_t^*) + u_t$, where u_t is a random disturbance with properties such that the variance of S_t is not a function of P_t^*, the covariance effect in (2.11) vanishes and $P_t^* = E[P_t]$.

Before exploring the second approach to studying anticipations, let us return to the nature of the producers' supply function, and the disposition of costs associated with risk and risk aversion. Let the cost function exclusive of costs associated with risk be $c(Q^*)$, and the corresponding function in which risk costs are included be $c_r(Q^*)$, so that $dc_r(Q^*)/dQ^* = P^*$. Let the producers' utility function for profit be $U( )$. The result in (2.11) comes from maximizing under competitive assumptions the implicit objective function, $E(PQ) - c_r(Q^*)$, which leads us in (2.9) and (2.10) to $P^* = E[R]/\mu$. A more typical specification in terms of utility is that producers maximize $E[U(PQ - c(Q^*))]$ with respect to planned output so that 'riskless' marginal cost $c' = E[U'P\varepsilon/\mu]/E[U']$, which is a transformation of the revenue expectation involving marginal utility. Derivation of the implicit objective function from the typical specification is only straightforward under either the trivial case of risk neutrality (when marginal utility is constant), or special cases when the utility function is separable into a linear 'expected' part, and a 'risk cost' part such as

$$(2.12) \qquad U(\cdot) = E[\cdot] - \phi V[\cdot].$$

Such a utility specification can be arrived at by arbitrary assumption, by assumption of normally distributed revenue and constant absolute risk aversion, or by an approximate Taylor-series expansion of expected utility.

The second approach to reviewing the nature of producers' anticipation is from the perspective of social efficiency. The question can be posed of what P* should be dictated in order to maximize expected welfare as measured in (2.8). Equating the derivative of E[W] with respect to anticipated price to zero for the first-order condition for maximal welfare gives, with permissible differentiation within the expectation operator[2]

$$(2.13) \qquad \partial E[W]/\partial P_t^* = E\left[(dP_t/dP_t^*)\epsilon_t g(P_t^*) + P_t \epsilon_t g'(P_t^*)\right]$$
$$- \mu E[g(P_t^*) + P_t^* g'(P_t^*)]$$
$$+ \mu E[g(P_t^*)]$$
$$- E[f(P_t)dP_t/dP_t^*]$$
$$= 0.$$

The right hand side of (2.13) can be simplified by introducing the market clearing condition from (2.4), that is, $f(P_t) = \epsilon_t g(P_t^*)$ whence

$$(2.14a) \qquad \partial E[W]/\partial P_t^* = E[P_t \epsilon_t g'(P_t^*)] - E[\mu P_t^* g'(P_t^*)]$$
$$= 0$$

which, on cancelling the common factor $g'(P_t^*)$ on the left hand side and noting that, in the differentiation, P_t^* is a decision variable, can be written as

$$(2.14b) \qquad E[P_t\epsilon_t - \mu P_t^*] = E[P_t \epsilon_t] - \mu P_t^*$$
$$= 0.$$

Since these expected values will be constant over time in a convergent market, the time subscripts can be dropped in the finally rearranged result of

$$(2.15) \qquad P^* = P_t^*$$

$$= E[P_t \, \varepsilon_t]/\mu$$

$$= E[P_t] + \text{cov}[P_t, \varepsilon_t/\mu]$$

$$= E[P] + \text{cov}[P, \varepsilon/\mu].$$

There is thus a happy and exact coincidence between the results of producers' enlightened self interests as expressed in (2.11), and the results of an altruistic dictatorship in (2.15) where welfare is measured by expected surplus. Although it is emphasized in (2.15), it is also implicit in the equivalent (2.11) that $E[P] = E[P_t]$ is the expected market clearing price given that producers indeed do plan on the basis of an anticipated certainty equivalent $P^* = P_t^*$ in each and every period. Comparing these alternative approaches to determining the nature of P^*, it is apparent that if producers anticipate expected returns in this market differently from the 'unit revenue' specification of (2.11), social welfare will not be maximal.

We conclude this section with a generalization of a further result from Hazell and Scandizzo that the rational price expectation P^* is lower than the self-fulfilling price expectation obtained when farmers make independent forecasts about prices and yields. The implication is that 'revenue' expectations lead to a smaller quantity being marketed on average than do 'price' expectations.

To review this topic let us reemphasize the conditionality of $E[P]$ in (2.15) on the assumed anticipations by writing for the moment, this price as $E[P|P^*]$. Now let us define a self-fulfilling price anticipation, $\tilde{P}$, as the expected market clearing price obtained when producers actually expect that price, that is, define $\tilde{P} = E[P|E[P]] = E[P|\tilde{P}]$. Demonstration that $P^* < \tilde{P}$ is by means of contradiction. For any anticipated price, $\tilde{P}$, expected market clearing price is obtained by equating supply and demand so that $E[P|\tilde{P}] = E[F(\varepsilon g(\tilde{P}))]$ where F is the inverse demand curve with $F' < 0$. Since $\partial E[\varepsilon g(\tilde{P})]/\partial \tilde{P} > 0$ then $\partial E[P|\tilde{P}]/\partial P < 0$. Supposing that $P^* > \tilde{P}$, $E[P|P^*] < E[P|\tilde{P}] = \tilde{P}$. Since by assumption $P^* > \tilde{P}$, then it follows that $P^* > E[P|P^*]$. However, this contradicts the finding in (2.15) where the covariance term is negative, as it must be with $F' < 0$, so that the supposition that $P^* \geq \tilde{P}$ is false, and the required result follows that $P^* < \tilde{P}$.

Long-Run Analysis

In spite of our assumptions about convergence of

markets and measurement of welfare by means of expectations, our analysis so far has been based on short-run
market equilibria in which producers realize, on average, a nonzero profit. The question naturally arises as
to what is the optimal anticipated price consistent with
zero profits in the long run. It was assumed in modeling supply that producers act on the basis of anticipated profits. However, their welfare depends on
expected realized (ex post) profits and so, in a long-
run equilibrium, it is required that both measures of
profit be zero. Only then will the number of firms
stabilize because ex ante and ex post expectations will
coincide.

Since producers are assumed to equate marginal cost
and anticipated price, ex ante pure profits will always
be zero in a long-run equilibrium. For long-run stability, expected ex post profit is also zero, or equivalently, long-run marginal and average cost (again
represented by P^*) are equated to long-run expected
marginal return. The relevant zero profit equation can
be written as expected return less expected cost as in

$$(2.16) \qquad E[P\varepsilon g(P^*)] - P^*\mu g(P^*) = 0,$$

and this can be simplified and expressed as $P^* = E[P\varepsilon]/\mu$. This is identical with the short-run results
because, as John Quiggin has pointed out to us, with
the multiplicative risk specification, marginal and
average revenues are everywhere equal, and the equality
is required for long-run equilibrium to hold.

A CASE OF CROSS-SECTIONAL VARIATION IN YIELDS

The result that producers should form expectations
as simple per unit revenues may seem to depend on an implicit assumption that yields are perfectly correlated
between producers. Suppose now that we enrich the assumption about yields and assume that k groups of producers (for instance in several regions) confront yields
that are not perfectly correlated with each other or, a
fortiori, with total yield.

The market model can be generalized for this case by
adding to (2.3) a subscript to denote supply region i,
and for simplicity, suppressing the time subscript as in

$$(2.17) \qquad S_i = \varepsilon_i g_i(P_i^*).$$

The demand specification is left unchanged, and the market clearing equation (2.4) becomes $D = \Sigma_i S_i$. Applying

16

the definitions of consumers' and producers' surplus in (2.6) and (2.7), expected welfare can be written as

(2.18)
$$E[W] = \Sigma_i\{E[P\varepsilon_i g_i(P_i^*)] - \mu_i E[P_i^* g_i(P_i^*)]$$
$$+ \mu_i E[\int_{P_s=0}^{P_i^*} g_i(P)dP\} + E[\int_{P}^{P_d=0} f(P)dP].$$

The welfare maximizing set of P_i^* are then found by simultaneously equating the k partial derivatives to zero, and the representative ith such equation is

(2.19)
$$\partial E[W]/\partial P_i^* = E[(\partial P/\partial P_i^*)\Sigma_i\{\varepsilon_i g_i(P_i^*)\}$$
$$+ P\varepsilon_i g_i'(P_i^*)] - E[\mu_i g_i(P_i^*)$$
$$- \mu_i P_i^* g_i'(P_i^*)] + E[\mu_i g_i(P_i^*]$$
$$- E[f(P)dP/dP_i^*]$$
$$= 0.$$

With the use of the market clearing condition and the cancellation of terms, this equation only contains the ith of the anticipated prices and so it can be solved independently as

(2.20)
$$\partial E[W]/\partial P_i^* = E[P\varepsilon_i g_i'(P_i^*)] - \mu_i P_i^* g_i'(P_i^*)$$
$$= E[P\varepsilon_i]/\mu_i - P_i^*$$
$$= 0.$$

Not unexpectedly, the rational price expectation or optimal anticipation for the ith group of producers is their respective standardized expected unit revenue (as it was for the single group of producers in (2.15)) which, as in (2.11), is also their expected marginal revenue.

Thus producers need not consider cross-sectional correlations among yields, and need only form revenue expectations based on their own yield expectations and the market clearing price. The corollary is that rational price expectations will naturally differ with yields between groups and that no single price expectation can be held rationally by all groups when such diversity prevails.[3]

A PARTIAL EQUILIBRIUM ANALYSIS OF A CASE OF
JOINT PRODUCTION

A question similar to that concerning the effect of cross-sectional yield variations on rational price expectations arises when joint production of risky crops is considered. Do yield correlations between crops have any effect on the rational price expectation for each market?

Consider a market context of two products where supplies are interdependent (joint production) but demands are assumed to be independent. The system may be specified as follows, where subscripts denote products 1 and 2, respectively:

$$S_1 = \varepsilon_1 g_1(P_1^*, P_2^*) \quad , \quad S_2 = \varepsilon_2 g_2(P_1^*, P_2^*)$$
$$D_1 = f_1(P_1) \quad \bullet \quad , \quad D_2 = f_2(P_2)$$

and the market clearing conditions are $D_i = S_i$, $i = 1,2$.

Using the introduced definition of producers' and consumers' surplus, the expected value of social welfare can be written as

(2.21)
$$E[W] = E[P_1 \varepsilon_1 g_1(P_1^*, P_2^*)] + E[P_2 \varepsilon_2 g_2(P_1^*, P_2^*)]$$

$$- \mu_1 E[P_1^* g_1(P_1^*, P_2^*)] - \mu_2 E[P_2^* g_2(P_1^*, P_2^*)]$$

$$+ \mu_1 E[\int_{P_{1s}=0}^{P_1^*} g_1(P, P_2^*) dP]$$

$$+ \mu_2 E[\int_{P_{2s}=0}^{P_2^*} g_2(P_1^*, P) dP]$$

$$+ E[\int_{P_1}^{P_{1d}=0} f_1(P) dP]$$

$$+ E[\int_{P_2}^{P_{2d}=0} f_2(P) dP].$$

Socially optimal levels of P_1^* and P_2^* are then found
using the procedures elaborated in Equations (2.13) and
(2.19) of taking partial derivatives with respect to
these two decision variables, setting these to zero for
the first-order optimal condition, employing the market
clearing assumptions and simplifying to yield from
(2.21) the equations

(2.22a)

$$\partial E[W]/\partial P_1^* = E[\{\partial g_1(P_1^*, P_2^*)/\partial P_1^*\}$$

$$\{P_1\epsilon_1 - P_1^*\mu_1\}]$$

$$+ E[\{\partial g_2(P_1^*, P_2^*)/\partial P_1^*\}$$

$$\{P_2\epsilon_2 - P_2^*\mu_2\}]$$

$$= 0.$$

(2.22b)

$$\partial E[W]/\partial P_2^* = E[\{\partial g_1(P_1^*, P_2^*)/\partial P_2^*\}$$

$$\{P_1\epsilon_1 - P_1^*\mu_1\}]$$

$$+ E[\{\partial g_2(P_1^*, P_2^*)/\partial P_2^*\}$$

$$\{P_2\epsilon_2 - P_2^*\mu_2\}]$$

$$= 0.$$

Solving these simultaneously yields the now standard
result that

(2.23a) $$P_1^* = E[P_1\epsilon_1]/\mu_1,$$

(2.23b) $$P_2^* = E[P_2\epsilon_2]/\mu_2,$$

and so optimal anticipations should be in terms of standardized expected unit revenues even in the specified
case of joint production. This result obviously gener-
alizes to the case of k products when the markets for
jointly produced commodities are as specified above. In
Chapter 4 we return to the specification of multiple
products in the context of mathematical programming
models in which more complex demand structures are
assumed.

GENERAL EQUILIBRIUM ANALYSES

One of the consequences of revenue expectations is that they lead, on average, to market prices and outputs different from those found with price expectations where no account is taken of yield variations. One possible defect of using partial equilibrium analyses may be the deliberate ignorance of wage and income effects in an economy. These effects might be expected to be important in developing countries where production is oriented to agricultural products generally and where food is an important component of household expenditure. We begin our general equilibrium analysis by considering a simple two-sector model in which wage and income effects can be studied and where producers' supply decisions depend on revenue or price anticipations. We then go on to a more general type of general equilibrium analysis where we discard the assumption that the marginal utility of income is constant and find that results concerning optimal anticipations are correspondingly complicated.

A Two-Sector General Equilibrium Model

For our short-run analysis of wage and income effects we adopt a considerably simplified two-sector model of Kelley, Williamson and Cheetham (1972). The two goods are produced at levels Q_1 and Q_2 with the help of two primary factors, labor, X, and capital. The goods Q_1 and Q_2 might be thought of as food and non-food, respectively. The national labor market is assumed to be perfect with full employment at an endogenous wage rate, p_X, whereas capital is assumed to be immobile between sectors and allocated exogenously. There are no savings or investment.

Production in the two sectors is described by two highly specialized diminishing-returns constant-elasticity functions

$$Q_1 = a_1 \varepsilon X_1^{0.5} \qquad \text{and} \qquad Q_2 = a_2 X_2^{0.5}$$

in which both labor elasticities are arbitrarily specified as one half in order to simplify greatly the following analysis. Only the production of Q_1 (food) is assumed to be risky, through inclusion of the multiplicative yield term ε which has mean μ and variance σ^2.

Again for simplicity, producers are assumed to maximize expected profits. With a varied specification of Q_1, as foreshadowed in equation (2.12), the subsequent results could also be obtained after allowing for risk-averse producers to maximize expected utility. In the case of the riskless non-food sector, we assume that the product price P_2 and the wage rate p_x are certain and known. This means that profit, which is given by $P_2 a_2 X_2^{0.5} - p_x X_2$, is also certain and known. Optimizing this expression with respect to the labor allocated to non-food production (X_2) gives an amount of labor, $\hat{X}_2 = (0.5 a_2/p_x)^2 P_2^2$ and, substituting this back into the production function gives the linear sector-supply function, $S_2 = (0.5 a_2^2/p_x) P_2$.

In the case of food production we assume inflexible allocation or lagged supply response, with producers acting on the basis of anticipated price P_1^* and a mean yield $E[\varepsilon] = \mu$. Their expected or ex ante profit is given by $P_1^* a_1 \mu X_1^{0.5} - p_x X_1$ and the analogous optimal labor allocation is $\hat{X}_1 = (0.5 a_1 \mu/p_x)^2 P_1^{*2}$. Expected supply is then $S_1^* = (0.5 a_1^2 \mu^2/p_x) P_1^*$ and actual supply is found by substituting the optimal labor allocation into the realized production function and is $S_1 = (0.5 a_1^2 \mu/ p_x) \varepsilon P_1^*$.

Income in the economy comprises total wage payments $p_x X$ (where $X = X_1 + X_2$ is the available supply of fully employed labor) and the sum of producers' surplus, which in this case represents rewards to capital. Since labor-use decisions are based on ex ante profits, the wage rate p_x and total labor payments $p_x X$ are nonstochastic. The producers' surplus in the non-food sector is $P_2 S_2 - p_x X_2$ and under our specific parameterization can be shown to be equal to $p_x X_2$. Ex ante profits in the food sector can analagously be shown to be equal to $p_x X_1$ but, because farmers act on the basis of anticipated price and yield, the ex post surplus is $p_x X_1 - (P_1^* S_1^* - P_1 S_1)$ where the bracketed term measures the discrepancy between anticipated and realized revenue.

Collecting terms, total realized income in the economy is

$$(2.24) \qquad Y = p_x X + p_x X_2 + p_x X_1 - (P_1^* S_1^* - P_1 S_1),$$

$$= 2 p_x X + P_1 S_1 - P_1^* S_1^*.$$

To describe the demand side of the economy simply we assume that laborers and entrepreneurs have the same simplified type of Stone-Geary utility function, so that demand comprises a linear expenditure system, $D_i P_i = \alpha_i Y$, $i = 1,2$, where the α_i are the marginal budget shares, and $\alpha_1 + \alpha_2 = 1$. The market clearing conditions, $D_i = S_i$, $i = 1,2$, complete our assumptions about this economy and enable the supply and income functions to be substituted into the demand functions to give the two-equation system:

(2.25) $\qquad (0.5a_1^2\mu/p_x)\epsilon P_1^* P_1 = \alpha_1\{2p_x X$

$\qquad\qquad\qquad + (0.5a_1^2\mu/p_x)\epsilon P_1^* P_1$

$\qquad\qquad\qquad - (0.5a_1^2\mu^2/p_x)P_1^{*2}\},$

(2.26) $\qquad (0.5a_2^2/p_x)P_2^2 = \alpha_2\{2p_x X + (0.5a_1^2\mu/p_x)\epsilon P_1^* P_1$

$\qquad\qquad\qquad - (0.5a_1^2\mu^2/p_x)P_1^{*2}\}.$

There are three prices in this system, P_1, P_2 and p_x, and, without further loss of generality, one (we chose p_x) can be taken as a numeraire (that is, the prices are scaled so that $p_x = 1$) so that equations (2.25) and (2.26) can be appropriately simplified. Solving (2.25) for P_1 gives

(2.27) $\qquad P_1 = \alpha_1(4X - a_1^2\mu^2 P_1^{*2})/(\alpha_2 a_1^2\mu\epsilon P_1^*),$

which, after substitution into (2.26) gives

(2.28) $\qquad P_2 = (4X - a_1^2\mu^2 P_1^{*2})^{0.5}/a_2.$

Equations (2.27) and (2.28) define the equilibrium market prices (in wage units) for a given realization of ϵ and a fixed value of P_1^*. Note that because of the simplifying assumption of the model, only P_1 is dependent on ϵ; P_2 is not stochastic and thus takes the same value regardless of the value of ϵ. An appropriate measure of the equilibrium value of P_1 is the expectation $E[P_1]$. This cannot be derived precisely without a more detailed stochastic specification of ϵ but can be approximated by using a second-order Taylor series approximation for the expected value of an inverse of a positive random variable, in which case

(2.29) $\qquad E[P_1] \doteq (1 + C[\epsilon]^2)\alpha_1(4X - a_1^2\mu^2 P_1^{*2})/$

$\qquad\qquad \alpha_2 a_1^2\mu^2 P_1^*),$

where $C[\epsilon] = \sigma/\mu$ is the coefficient of variation of ϵ.

We now turn to the question of social efficiency and thus to the derivation of the rational price expectation of the model, that is P_1^*. The appropriate welfare function to use here is the underlying Stone-Geary utility function, $W = \alpha_1 \log S_1 + \alpha_2 \log S_2$, on which the linear expenditure system is based. However, since S_1 is stochastic, we need to take the expected value of W. Using the realized supply functions and substituting the result in (2.28), the welfare function can be written as

$$(2.30) \qquad E[W] = \alpha_1 E[\log(0.5a_1^2 \mu \varepsilon P_1^*)]$$

$$+ \alpha_2 \log(0.5a_2(4X - a_1^2 \mu^2 P_1^{*2})^{0.5}).$$

Resorting again to a second-order Taylor series approximation,

$$E[\log(0.5a_1^2 \mu \varepsilon P_1^*] \doteq \log(0.5a_1^2 \mu^2 P_1^*)$$

$$- 0.5C[\varepsilon]^2.$$

Substituting into (2.30) and maximizing with respect to P_1^*,

$$\partial E[W]/\partial P_1^* \doteq \alpha_1/P_1^* - \alpha_2 a_1^2 \mu^2 P_1^*/$$

$$(4X - a_1^2 \mu^2 P_1^{*2})$$

$$= 0,$$

which can be simplified and solved for the socially optimal anticipation

$$(2.31) \qquad P_1^* = (4\alpha_1 X/(a_1^2 \mu^2))^{0.5}.$$

This rational expectation has two interesting properties, the first of which being that P_1^* is yet again a unit revenue expectation. To see this, consider the unit revenue expectation

$$\tilde{P}_1 = E[P_1 \varepsilon]/\mu,$$

and, substituting for P_1 from (2.27),

$$\tilde{P}_1 = E[\alpha_1(4X - a_1^2\mu^2\tilde{P}_1^2)/(\alpha_2\mu^2 a_1^2\tilde{P}_1)],$$

which can be solved for $\tilde{P}_1$ as

$$\tilde{P}_1^2 = 4\alpha_1 X/(a_1^2\mu^2),$$

which is equivalent to P_1^* of Equation (2.31). Farmers in this modeled world apparently do not need to have direct knowledge of the parameters in Equation (2.31) to form rational expectations but need only anticipate the average of their own revenues and yields.

The second property to be noted is the relationship between P_1^* and the self-fulfilling price expectation $E[P_1]$ that would arise if farmers formed forecasts about prices and yields independently. If they did use $E[P_1]$ as their price forecast, $E[P_1]$ can be substituted for P_1^* in (2.29) and the equation solved for $E[P_1]$ giving

(2.32a) $\qquad E[P_1] \doteq (4\alpha_1 X/(a_1^2\mu^2))^{0.5}\{(1 + C[\epsilon]^2)/$

$$(1 + \alpha_1 C[\epsilon]^2)\}^{0.5}$$

or

(2.32b) $\qquad E[P_1] = P_1^* k^{0.5}$

where P_1^* is the rational expectation in (2.31) and $k = (1 + C[\epsilon]^2)/(1+\alpha_1 C[\epsilon]^2)$ is a constant. Since $0 < \alpha_1 < 1$, then $k > 1$ and $E[P_1] > P_1^*$.

In words, the approximate rational expectation of this model implies a lower market output than for the approximate simple price expectations equilibrium, despite the wage or income effects in the rest of the economy. Further, the extent of production cutback depends on α_1, the marginal expenditure on food, and on $C[\epsilon]$. The more that food is important in national household expenditure, then the smaller the cutback should be in production. This result supports the notion that the Hazell-Scandizzo argument for distorting risky markets in which prices rather than revenues are anticipated is potentially more important in developed countries where α_1 is relatively small, but that such a distortion effect would be less important in less developed countries where α_1 is large. From the expression for k it is also apparent that, for any α_1, the effect will increase with riskiness of yields as measured by $C[\epsilon]$, the coefficient of variation of yields.

A General Equilibrium Model with Nonlinear Utility

So far in our analysis, we have consistently assumed that the marginal utility of money is constant. We turn now to a general equilibrium model in which this assumption is relaxed.

Again to simplify our analysis we assume a static and closed economy in which there are no savings, investment or international trade.

Production in the economy is assumed to be risky, and the $k \times 1$ vector of realized outputs is $Q = h(P^*, u)$ where P^* is a $k \times 1$ vector of anticipated or certainty equivalent prices, and u is a $k \times 1$ vector of stochastic disturbances with expected value $E[u] = \mu$. Producers plan on the basis of their anticipated prices and their associated supply, $Q^* = E[h(P^*, u)]$, and are confronted by a production possibilities frontier that can be described by a transformation function $T(Q^*) = 0$.

Producers are assumed to maximize certainty equivalent profits (or equivalently, their expected utility) subject to the transformation function, so that first-order conditions for optimizing are obtainable from the Lagrangian, $L_1 = P^{*t}Q^* + \lambda_1 T(Q^*)$, where P^t here denotes the transpose of P. The conditions are

(2.33) $$\partial L_1/\partial Q^* = P^* + \lambda_1 \partial T(Q^*)/\partial Q^*$$

$$= 0,$$

where

$$\lambda_1 = -\partial(P^{*t}Q^*)/\partial T(Q^*).$$

We assume that consumers have identical preferences which can be expressed by a utility function, $U(D)$, where D is a $k \times 1$ vector of goods consumed. Consumers are also confronted by an income constraint, $P^tD = Y$, where P is a $k \times 1$ vector of market clearing prices, and Y is total income. Forming the Lagrangian, $L_2 = U(D) - \lambda_2(P^tD - Y)$, the first-order conditions for optimality are

(2.34) $$\partial L_2/\partial D = \partial U(D)/\partial D - \lambda_2 P$$

$$= 0,$$

where $\lambda_2 = \partial U(D)/\partial Y = U'$ is the marginal utility of income.

Consider now the derivation of the rational expectation of P^* for this model. The appropriate problem can be formulated as the maximization of consumers' expected utility $E[U(D)]$, with respect to P^* such that $T(Q^*) = 0$ and $D = Q$. The Lagrangian of this problem is $L_3 =$

$E[U(Q)] + \lambda_3 T(Q^*)$, so that the first-order conditions for a maximum for each P_i^*, are, for $i = 1, \ldots, k$,

$$(2.35) \qquad \lambda L_3/\partial P_i^* = E[(\partial U/\partial Q_i)(\partial Q_i/\partial P_i^*)]$$
$$+ \lambda_3 (\partial T(Q^*)/\partial Q_i^*)(\partial Q_i^*/\partial P_i^*)$$
$$= 0,$$

where

$$\lambda_3 = -\partial E[U]/\partial T(Q^*).$$

Equation (2.33) can be specialized and rearranged as $\partial T(Q^*)/\partial Q_i^* = P_i^* \partial T(Q^*)/\partial (P^{*t}Q^*)$. Also, using the market clearing condition $Q = D$, (2.34) can be rearranged as $\partial U/\partial Q = U'P$ and specialized to $\partial U/\partial Q_i = U'P_i$.

Substituting these expressions into (2.35) and simplifying, it becomes

$$(2.36) \qquad E[U'P_i \partial Q_i/\partial P_i^*] - U_e' P_i^* \partial Q_i^*/\partial P_i^* = 0,$$

where $U_e' = \partial E[U]/\partial (P^{*t}Q^*)$ is marginal expected utility evaluated at anticipated revenue P^*Q^*.

Solving (2.36) for P_i^* gives

$$(2.37) \qquad P_i^* = E[U'P_i \partial Q_i/\partial P_i^*]/(U_e' \partial Q_i^*/\partial P_i^*)$$
$$= E[U'P_i \partial Q_i/\partial Q_i^*]/U_e'.$$

For the special case when the production risk terms are linear and multiplicative so that $Q_i = u_i h_i(P_i^*)$, then (2.37) reduces to

$$(2.38) \qquad P_i^* = E[U'P_i u_i]/U_e' \mu_i),$$

which has a clear similarity to the result obtained for utility-maximizing producers derived in the preamble to Equation (2.12) where the left-hand side was riskless marginal cost. In (2.38), P^* is then seen to be a type of unit revenue expectation but one in which the unit returns are modified through the marginal utility of income. Not surprisingly, if marginal utility is assumed to be constant, then (2.38) reduces to the earlier result, $P_i^* = E[P_i u_i]/\mu_i$, that is, where the optimal anticipated price is a standardized revenue expectation. Thus, while the revenue expectation is the rational expectation under some specializations of an economy, this finding does not hold in general.

A further insight to Equation (2.37) is found by returning to producers' behavior as modeled in (2.33) and, using a suggestion of John Quiggin, assuming that producers maximize expected utility of returns subject to the same transformation function. Suppose that their utility functions are identical for all producers and can be represented aggregatively as $U_p(P^tQ)$. Then the Lagrangian is $L_4 = E[U_p(P^tQ)] + \lambda_4 T(Q^*)$ and the first-order conditions can be represented by

$$(2.39) \qquad \partial L_4/\partial Q_i^* = E[(\partial U_p/\partial(P^tQ))P_i \partial Q_i/\partial Q_i^*]$$

$$+ \lambda_4 \partial T(Q^*)/\partial Q_i$$

$$= 0,$$

where $\lambda_4 = \partial E[U_p(P^tQ)]/\partial T(Q^*)$. Equation (2.33), which is equivalent to the maximization of L_4 if P^* is a vector of certainty equivalent prices, specializes to $\partial T(Q^*)/\partial Q_i = -P_i^*/\lambda_1$. Substituting this into (2.39) and rearranging it explicitly for P^* gives

$$(2.40) \qquad P_i^* = (\lambda_1/\lambda_4)E[(\partial U_p/\partial(P^tQ))P_i \partial Q_i/\partial Q_i^*].$$

Writing $\partial U_p/\partial(P^tQ) = U_p'$, the marginal utility of producers' gross income, and substituting for λ_1 (from 2.33) and λ_4 (from 2.39) in (2.40) and cancelling $\partial T(Q^*)$ gives

$$(2.41a) \qquad P_i^* = E[U_p' P_i \partial Q_i/\partial Q_i^*]/(\partial E[U_p(P^tQ)]/\partial(P^{*t}Q^*)),$$

or

$$(2.41b) \qquad P_i^* = E[U_p' P_i \partial Q_i/\partial Q_i^*]/U_{pe}',$$

where $\partial E[U_p(P^tQ)]/\partial(P^{*t}Q^*) = U_{pe}'$ is producers' marginal utility evaluated at their anticipated total revenue.

In comparing the results of (2.37) and (2.41b) it is evident that they are similar and, indeed, are identical if producers have utility functions with properties identical to those of the functions assumed for consumers, i.e. $U' = U_p'$, in which case $U_e' = U_{pe}'$. This will occur, of course, when the producers are also the consumers and when risks are spread between the two groups through the prevalence of a perfect capital market. Otherwise, producers' anticipatory behavior will tend to diverge from the rational expectations behavior, and intervention may be desirable to increase consumers' welfare.

CONCLUSION TO THE THEORETICAL WORK

Only a few simple models from an infinite set of conceivable models of risky markets have been considered in this chapter. However, in all the several models reviewed, the relevance of producers' contemplation of both prices and yields together in forming their anticipations about risky product prices has been established and seems likely to be a quite general result. In partial equilibrium models of specifications that seem plausible from an empirical point of view, the result is that producers should, in their own and society's best interests, anticipate standardized unit revenues rather than merely prices. The findings in general equilibrium settings naturally tend to be more complex and add to the simple anticipation of unit revenue features of utility functions of groups within society. Thus the simple results should be applied carefully in any normative applications (such as project appraisals) where such considerations may be important-- for example, in developing countries where income is low, marginal expenditure on food is high and utility of income is probably markedly concave (risk averse).

However, we cannot escape the central conclusion that, when production is risky and competitive producers face downward sloping demand curves, producers should plan their production on the basis of expected revenues. If they are averse to risk, their anticipatory and planning tasks are more challenging but, under some particular attitudes towards risk, still reduce exactly to anticipatory expected revenue. As a good working rule, such anticipations will always approximate closely those that are optimal even under more complex representations of producers' maximizing behavior.

NOTES

1. This follows from the general result that if x and $f(x)$ are random, then the covariance (and correlation) between x and $f(x)$ is $>$ or < 0 depending on whether the derivative $f'(x) >$ or < 0. For example, using a proof due to Richard Just, if $E(x) = \bar{x}$ and $f'(x) < 0$, then $f(x) > f(\bar{x})$ if $x < \bar{x}$ and $f(x) < f(\bar{x})$ if $x > \bar{x}$. It follows that $(f(x) - f(\bar{x}))(x - \bar{x}) < 0$ for all $x \neq \bar{x}$ in the sample space, and cov $[f(x),x] = E\{[f(x) - f(\bar{x})](x-\bar{x})\} < 0$, unless the distribution is singular at $\bar{x}$.

2. For the convenience of readers who wish to follow this derivation, the four lines correspond with the four terms in (2.8). The first two lines are derivatives of products. The third involves a derivative of a definite

integral of which the lower bound is zero (supply at $P_{s=0}$). The fourth involves a similar case, but where the upper bound is zero (demand at $P_{d=0}$), hence the negative sign, and the function-of-a-function rule must be invoked.

3. Newbery and Stiglitz (1981, p.138) arrive at a similar result. They assume that yield risks can be decomposed into a general yield risk common to all farmers, and therefore correlated with price, and a component which is specific to the individual farmer or region and which is not correlated with price.

3
Empirical Contrasts of Competing Arguments of Agricultural Supply Functions

In Chapter 2, the issue of whether producers' expectations in determining how much to produce are formulated in terms of merely prices or, more desirably, unit revenues, or perhaps something else, has been raised through several specific theoretical considerations. In this chapter we approach the question of anticipated price versus revenue empirically by confronting time-series data for aggregate agricultural supplies. The empirical work is prefaced by a brief review of some further theoretical topics.

FURTHER THEORETICAL CONSIDERATIONS

Aspects of the relevant literature of production economics can be examined briefly under three categories --models of the firm (by way of background), aggregate models, and models that include risk aversion.

Models of the Firm

Firms typically, and farm firms especially, operate in an environment where their productivities and the prices they face are risky. It is thus surprising that modelers of the firm have devoted scant explicit attention to the effects of jointly risky prices and yields. When these effects have been considered it has usually been in the context of maximizing models of the problem of optimal choice of inputs. It is found that optimal input levels, and thus also levels of planned output, should in general depend on any statistical nonindependence between prices and yields. Even in the extreme case of a simplifying assumption of neutrality of attitude towards risk, the expression for the best bundle of inputs involves the covariance (or correlation) between output price and productivity (Magnússon 1969, p. 104;

29

Anderson, Dillon and Hardaker 1977 , p. 172). The covariance effect is absented only be presumption that it
is zero. The effect was also always evident from the
slightly different perspective of the output-oriented
production models described in Chapter 2.

Virtually no empirical effort has been addressed to
modeling such correlation effects, although some minor
attempts are represented by Anderson (1974, pp. 176-178;
1977). The neglect might be understood by contemplating
the opaqueness of some of the theoretical results and by
observing that perceptive analysts have approached the
problem implicitly. They have sidestepped the difficulties of dealing explicitly with price-yield correlations
by working simply with (observed) revenues which effort-
lessly embody any joint effects. This, for instance,
has been the method typically adopted by linear (for
example, Hazell 1971a) and quadratic (Freund 1956) risk
programmers in models of multi-enterprise firms.

To the extent that risky prices and yields confronted by individuals are not stochastically independent, output decisions thus properly depend on the joint
distribution. Firm-level modelers employing risky revenues have at least implicitly recognized this desider-
atum. An alternative statement of the conclusion about
individual supply decisions is that, in general and even
under risk neutrality, supply should be a function of
the expected value of price and the covariance of price
and yield. Whether this translates into market supply
functions with arguments of revenues rather than prices
depends on the extent to which producers attempt to maximize the objective functions discussed above, the perceptions of covariances by individuals and, relatedly,
the relationship between individually perceived and
market-wide covariances.

Aggregate Supply Analysis

Aggregate analysis of the supply schedules for agricultural commodities has proved a very popular pastime
for agricultural economists, as witnessed by the pleth-
ora of studies reviewed, for example, by Seagraves
(1970), Anderson (1974), Askari and Cummings (1976), and
Tomek and Robinson (1977). With few exceptions, the
analysts have exploited the Nerlove (1956, 1958) model
of supply response via distributed lags where producers
are recognized as having to make decisions without complete knowledge of the prices they will receive. Again
with very few exceptions, these many analysts have postulated an adaptive expectations model predicated on
adjustments to past prices.

Against this weight of circumstantial evidence consistent with the suggestion from elementary economics

that price is indeed the conventional and compelling argument of a supply function, stand a few not-so-recent studies in which the analysts have related supply to revenue (per unit area). Three of these, namely Sengupta and Sen (1969), Just (1974), and Castro and Seagraves (1974), are singled out initially for comment.

To be sure, the motives of these authors were not entangled with the question of alternative expectations models posed by the contrasting results of Hazell and Scandizzo (1975, 1977). Castro and Seagraves recognized that they were dealing with a dynamic agriculture with non-stationary yields (of Florida winter green peppers). In fact, their handling of revenues is by means of a lagged price structure times a normal yield expectation, and thus abstracts from any correlations existing between prices and yields. Just (1974, p. 55), in his study of supply response of 38 California field crops, used revenues as a 'desirable . . . simplification' in place of using both prices and yields as explanatory variables.

Sengupta and Sen (1969) appear to be the only non-recent analysts to have made a systematic comparison be-tween prices and unit revenues (which they term returns) in supply functions (for rice and jute). They noted (p.8) that '. . . since "yield" is largely uncertain, the supply response (or acreage response) of the farmer will depend on the mean and variance of "anticipated" return rather than those of prices alone . . . Thus . . . on "a priori" grounds or on the basis of our data characteristics there is no definite case for preferring one model specification as "the best".' While emphasiz-ing the tentative nature of their results they found, statistically speaking (p.21), that the lagged price variables performed better than the revenue variables in explaining supply response effects beyond the influence of contemporaneous area-sown effects. Neither lagged variable was very statistically influential in determin-ing areas sown (p.23) although the signs of the price effects were more theoretically consistent than were those for revenue effects.

Analogous results were found by Griffiths and Anderson (1978) in a regional study of wheat area response in Australia, inspired by the theoretical work of Hazell and Scandizzo (1977). They were surprised to find in their case study that anticipations about prices alone seemed to explain very slightly more variance in time series of areas sown than did anticipations about unit revenues. The incredulity in their findings was heightened by their observations that risk-responsive-ness was not statistically significant and that the random variation in functions for quantity supplied was represented better as an additive rather than a multi-plicative disturbance. Subsequent work by Sanderson,

Quilkey and Freebairn (1980) on similar Australian wheat industry data, but differentiated by the netting out of costs in the revenue contrast, was very similar in finding virtually no difference in the explanatory power of the price only versus revenue models of partial adjustment and adaptive expectations.

In summary, conventional econometric practice seems generally consistent with the notion that anticipated prices are of major importance. Perhaps though, a new fashion based on anticipated revenues is in the process of emerging.

Before departing discussion of supply at the aggregate level, it can be observed that analysts using programming or normative models for studying supply response typically have emphasized price anticipations and disregarded any price-yield correlations. Most of the latter-day aggregative programming models (see for example Duloy and Norton 1973, and Norton and Schiefer 1980), in which prices are endogenous, do not permit accounting for price-yield correlations (excepting in-so-far as these enter the variance or risk-aversion accounting). However, Hazell and Scandizzo (1977) have developed appropriate procedures and these are elaborated below in Chapter 4.

Risk Aversion in Supply Response

Not too much has been done in supply analysis about representing the risk faced by producers and any aversion to it that they might have, until the past decade or so. In microlevel models (of both the generalized and programming types noted above), risk aversion has been recognized as an important influence conditioning response. Empirical progress, however, has usually rested on measuring risk by the variance or standard deviation of revenues and, in turn, this implies the prevalence of either rather restrictive families of risk-averse utility functions, or rather particular types of probability distributions (notably the normal).

The micro-implications for aggregate response have been tracked most successfully in mean-standard deviation linearized aggregative programming models, such as those of Hazell and Scandizzo (1974, 1977). Matching work of econometric style has been virtually nonexistent (an intermediate case is the 'portfolio' model of Sengupta and Sen (1969)) and seems represented mainly by the supply studies of Behrman (1968) and Just (1974), (the latter being variously refined and summarized in Just (1975, 1976))[2]. In these studies, producers' risk-averse supply response is cast in a mean-variance framework. Behrman used mean and a three-year moving average variance of price while Just used measures of evolving subjective mean and variance of revenue. The latter measures are better rationalized and more general than Behrman's, and accordingly are the subject of further attention herein.

Summary of Review Material

Producers confronting uncertain prices and yields (and therefore revenues) must commit resources to risky productive activities on the basis of anticipations about the future. Intuition and tradition tell us that these anticipations must depend importantly on prior experience--but experience of what? Most analysts have presumed that past prices are of transcending importance and, further, that their importance is carried solely in the anticipated mean (expected) price. Contrary to the implication of convention, we suggest from our review of theory that past revenues--virtually untried by empiricists--may be more apposite.

Likewise, producers' responsiveness to changes in subjective risk has received scant attention, particularly in econometric work, in spite of wide intuitive acknowledgement of the probable importance of such a phenomenon of risk aversion. This issue ascends to potential prominence when stabilization schemes are contemplated (Newbery and Stiglitz 1981, chap. 22). Many results from Nerlovian (risk-neutral) adaptive expectations appear promising from a statistical point of view, yet prove disappointing in predictive applications. Just (1975) has argued that one explanation for this is the disregard of the risk effect which, in the estimation, tends to be carried in the estimate of the coefficient of expectation.

Empirical contrasts between the competing arguments of supply functions would thus seem to be called for if the issues raised in Chapter 2 are to be given real-world context and relevance. This is tackled here in two stages. First, a review is made of several case studies in which data from several specific countries are analyzed in various independent and conventional econometric models. Unfortunately, these case studies prove to be inconclusive for resolution of whether revenues or merely prices are dominantly operative as arguments of supply functions. They are presented here in abbreviated form to emphasize the empirical delicacy associated with resolving the question at hand. In search of resolution, the second stage is embarked upon wherein the topic is addressed with compatible sets of cross-country time-series data and an inferential setting of non-nested hypotheses. This second stage, in spite of necessarily abstracting from analysis of risk

responsiveness, is somewhat more definitive, and so we
proceed to our conclusions concerning supply analysis.

SOME CONVENTIONAL ECONOMETRIC CASE STUDIES

Several case studies were undertaken on an essentially ad hoc basis according to data conveniently to
hand at the time of active investigation. These are now
described very briefly and the limited results then
reviewed. Those to be sketched here employ data on
field-crop area responses to changing economic conditions in Mexico, South Asia, and the state of North
Carolina, USA. Some results for Australia from a study
by Griffiths and Anderson (1978) are also included in
the review. The focus on major nonperennial field crops
is deliberately restrictive. Analysis of supply
response for perennial and tree crops and most livestock
products is intrinsically more difficult and, because of
the capital good/inventory aspects of such products,
would be unlikely to be as revealing about arguments of
supply functions as would analysis of annual crop
phenomena.

Irrigated Field Crops in Guasave, Mexico

In the Guasave irrigation district of the Pacific
Northwest region of Mexico, approximately 17,000 hectares of irrigated (one-cycle) field crops are grown
annually. The CHAC aggregative programming model of
Mexican agriculture (Duloy and Norton 1973) included
thirteen crops dominant in this district. Choice of
this region for study was arbitrary and was primarily
based on the ease and feasibility of the intention of
using linear programming to investigate 'normative'
supply response in a contrasting study of the same
region. The latter, however, proved abortive because of
the inability to capture responsiveness to changing
historical prices with the existing model (including
risk-averse decision making as described by Hazell et
al. (1983)). Rather similar results to those overviewed
here were obtained in analogous studies of two other
Mexican irrigation districts (San Juan and La Laguna).
Only the Guasave results are discussed for the sake of
brevity and because the results are generally inconclusive.

Two analytical methods were used, namely the traditional adaptive expectations approach of Nerlove (1958)
and the more recently developed approach of Just (1974).
One of the reasons for the popularity of the Nerlovian
approach is that, with some abuse of the underlying
assumptions about the conceptual disturbance term,

ordinary least squares (OLS) regression can be used to furnish immediate (albeit biased and inconsistent) results. By contrast, maximum likelihood (ML) search and nonlinear regression methods are required to implement the Just approach.

It is simplistic to postulate only own-price effects for a world which obviously involves simultaneous and interactive determination of areas devoted to all the thirteen important crops. Attempts were made to capture cross-price effects without resorting to the type of whole-system estimation methods such as advocated (and illustrated for a three-enterprise case) by Powell and Gruen (1966a, 1966b, 1968). In the Nerlove models, a cross-price (or revenue) index was composed as a Laspeyres index of all prices other than for the crop being analysed, using mean crop areas as the quantity weights. In the Just models, variables for own-price mean and variance effects were supplemented by variables for the means and variances of up to three of the substitutable other crops' prices and revenues.

With thirteen crops, two estimational methods, a variety of possible specifications concerning effects to be allowed, and functional forms to be used, the results were voluminous and not always very satisfactory from a statistical viewpoint. Results for the six most minor crops were least satisfactory and attention was collapsed to the seven major crops--safflower, wheat, rice, maize, sorghum, cantaloupes and tomatoes.

While the estimated supply functions may be of inherent interest, the emphasis here is on the competition between prices (P) and revenues (R). One summary of the results is included in the first two rows of Table 3.2. It can be seen that the Just method proved superior to the Nerlove method, in part because of the evident prevalence of risk responsiveness, but that neither price nor revenue stood out as a generally superior argument.

This undefinitive result was also confirmed in another summary analysis. The procedure was adopted of representing each response function only by an overall statistic for goodness of fit, namely the coefficient of multiple determination. For each method of estimation the two sets of coefficients were taken as two samples of observations and arrayed as sample distribution functions. In no case did these differ significantly--by intuitive appraisal, by notions of stochastic dominance, or by the nonparametic Kolmogorov-Smirnov test of the null hypothesis that both samples arose from a common parent population. This case study was thus inconclusive for insight to the nature of price-yield effects in producers' anticipatory behavior.

Jute in Bangladesh and India

Estimation of supply functions for jute in major ex-
porting countries was undertaken as part of a study of
jute price stabilization. Aggregate supplies for each
of Bangladesh and India were considered. It was considered important to include responsiveness to changing
risk in the stabilization study. Accordingly, the Just
model was used. Yield was found to be unresponsive to
price and so only the area planted to jute was considered as the decision variable. This accords with the
procedure adopted in the other case studies.

The most important competitive crop in the jute-
growing areas is autumn rice. This competitive effect
was examined in various ways in preliminary analyses and
the mean rice price was found to be quite important in
influencing jute areas. However, variability of rice
price and covariability with jute were not found to be
influential in jute supply. Thus the selected model included subjective mean and variance of jute price (or
revenue) and mean rice price (or revenue).

Undeflated prices were available and, with recent
and prospective high rates of inflation in Bangladesh
and India, it seemed desirable to attempt to specify the
supply equations in such a way as to minimize the problems connected with ignored inflation in simulation of
stabilization policies. Thus the model was specified in
terms of responsiveness to the ratio of mean jute price
(revenue) to mean rice price (revenue) and to the dimensionless subjective coefficient of variation. A term
embodying the effect of variance prior to the observation period (1952/53 to 1974/75) was included. Data for
two years prior to the observation period were used for
estimation of prior means.

The results for the price-only specification are
summarized in Table 3.1. Attention can usefully be
concentrated on the first row where the summary
goodness-of-fit statistics are reported. The revenue
specification was inferior and unworthy of reporting.
At this high level of aggregation it apparently resulted
in poorer goodness of fit, some statistically insignificant coefficients, and estimates of the parameter for
subjective mean updating that were outside the theoretically acceptable range. In this instance, the revenue
specification seemingly merely introduces unwarranted
random effects into the estimation of producers'
response to price and does not assist in resolving the
question of how they take account of the joint price-
yield effects.

TABLE 3.1
Summary of Estimated Jute Supply Functions
for Bangladesh and India

	Bangladesh	India
Coefficient of determination[a], R^2	0.776 (0.519)	0.717 (0.525)
Durbin-Watson statistic	2.12	2.34
Standard error of the estimate[b]	89	73
Subjective expectation parameter	0.294 (0.096)[c]	0.793 (0.104)[c]
Subjective risk parameter	0.103 (0.044)	0.287 (0.077)
Intercept	323 (269)	422 (107)
Ratio of mean prices	651 (216)	275 (71)
Coefficient of variation	-1230 (373)	-992 (276)
Prior variance	-625 (159)	-22 (77)

[a] Numbers in parentheses in this row are for the corresponding function specified in terms of revenues.
[b] Units are 10^3 hectares.
[c] These and following bracketed standard errors are asymptotic.

Field Crops in North Carolina, U.S.A.

Fletcher and Gellatly (1977) estimated supply response functions of the Just (1974) type for several field crops in North Carolina, using state aggregate data. Their results, which involved formulating the price variables as unit revenues, were excellent for the four major crops; maize, cotton, wheat and soybeans. The authors kindly agreed to re-estimate their supply functions for these four crops in terms of prices.

The econometric procedure used involved simultaneous estimation of the effects of mean and variance prior to the period of observation. Each crop area was postulated to respond to its own mean and variance (of price or revenue), its own prior mean and variance, and to a set of policy variables representing various government interventions.

The results are summarized in Table 3.2. All the coefficients of multiple determination exceed 0.9 but there is no consistent pattern of superiority according to whether prices or unit revenues are used. These

TABLE 3.2
Summary of Some Case Studies of Supply Analyses Comparing Price
and Revenue Expectations

Data	Length of time series (years)	Model and estimational method[a]	Number of crops or regions	Cases with risk responsiveness	Comparisons[b] where price, P, or revenue, R, was strongly (weakly) favored or similar (=)				
					P	(P)	=	(R)	R
Mexico (Guasave District) Fieldcrops	13	N	7	–			2		2
	13	J	7	6	4		2		2
Bangladesh and India Jute	23	J	2	2	2				
North Carolina Fieldcrops	25	J	4	4				4	
Australia (NSW) Wheat	20	J	1	0		1			

a N denotes the Nerlove model of adaptive expectations estimated with ordinary least squares regression and J denotes the Just model of subjective mean and variance expectations estimated with maximum likelihood search.

b Cases where the results were of quality too inferior to warrant comparisons are excluded.

equivocal results unfortunately provide no guidance as to whether prices or revenues are more appropriate because there is nothing apparent as a basis for discrimination.

Overview

The main features of the case studies are drawn together in Table 3.2 which also includes the results of Griffiths and Anderson (1978) for wheat (area) supply response in southern New South Wales, Australia.

No clear pattern is apparent in the results shown in the right-hand side of Table 3.2.

The main conclusion, reinforced by similar findings for other Mexican regions, is that it is impossible to say from these conventional results whether producers are guided more by price or by revenue expectations in their crop planning decisions. Thus the question about the appropriate behavioral argument of risky supply functions remains unresolved--paralleling the experience of Reutlinger (1964).

Perhaps conventional econometrics is inappropriate for resolution of such an issue in an unequivocal manner? More direct approaches may be required. As Schaller (1969, p.368) observed, 'we have not really studied the relationship of producer plans to their expectations, nor the relationship between their plans and actual behavior.' Perhaps we should follow the advice given to him (p.368) that, if you want to know what farmers do, why don't you ask them? As researchers found by the early 1950s, it is not a trivial matter to model farmers' expectations processes,[3] and as Fisher and Munro (1983) observe, the process of gathering individuals' expectational data can be quite expensive, especially relative to historical aggregate data. It may be necessary to study the formulation of expectations and production plans by panels of farmers over quite a few years to hope to obtain clear resolution. Experimental methods such as developed by Schmalensee (1976), Fisher and Tanner (1978), and Fisher and Munro (1983) may prove useful but mind experiments alone do not encompass the linkage of expectations to real production planning.

However, before abandoning econometric approaches to understanding anticipations in supply functions, we take up our second stage of examination armed with new sets of data that enable pooling of information across regions and, eventually, with statistical procedures specifically designed to discriminate between the two models under consideration with appropriate recognition of their 'non-nested' nature.

SOME UNCONVENTIONAL ECONOMETRIC CASE STUDIES

Given the poor ability to distinguish among the complex risk models based on either prices or revenues demonstrated above, it seems that perhaps use of a simpler model might lead to an enhanced ability to distinguish between the alternatives. Accordingly, we resort only to cases where changes in risk are not considered in order to avoid the complications, the additional variables and the tendency towards multicollinearity associated with risk terms in the regression equation. Another simplification suggested by Just (1977) is to drop the prior mean (and variance terms) from the model used above as well. Just shows that the estimates of the remaining parameters maintain consistency with this simplification, whereas coefficients for the prior mean (and variance terms) cannot be estimated consistently with sample data anyway.

An additional modification of the alternatives used in the case studies above seems necessary for conclusive results to be forthcoming. That is to add yield expectations to the model in which price expectations appear. Thus, we turn to an investigation of alternative hypotheses where crop areas either depend on price expectations and yield expectations or on revenue expectations. The yield expectations variable is added to the former model since otherwise statistics might favor the revenue model simply because of the importance of yields rather than because the data manifest farmers' consideration of covariance between yields and prices.

A Simplified Model for Hypothesis Testing

Consider the alternative models

$$(3.1) \qquad A_t = a_0 + a_1 E_{pt} E_{\epsilon t} + w_t$$

versus

$$(3.2) \qquad A_t = b_0 + b_1 E_{rt} + w_t,$$

where A_t is crop area, a_i and b_i are parameters, the w_t are stochastic disturbances with $E[w_t] = 0$; and E_{rt}, E_{pt}, and $E_{\epsilon t}$ represent subjective expectations of revenue per unit of area, price per unit of output, and yield per unit of area, respectively. Note that price and yield expectations are combined multiplicatively in the former model to minimize the number of parameters and thus reduce the possibilities that collinearity can prevent model discrimination. Note also that a multiplicative combination of price and yield expectations is consistent with the treatment in the programming models discussed in Chapter 4.

Within the context of the above models, the expectation terms are specified as truncated geometric lags on the corresponding variables; for expected price, for example,

(3.3) $$E_{pt} = \sum_{i=1}^{k} d_i P_{t-i},$$

where $d_i = \theta^i \left[\sum_{j=1}^{k} \theta^j \right]^{-1}$ and where $0 < \theta < 1$.

Note that the factor $\left[\sum_{j=1}^{k} \theta^j \right]^{-1}$ is used to inflate the weights of the truncated geometric distribution so that they sum to unity.

In the context of the above model, a nested hypothesis can be constructed by simply rewriting Equation (3.2) as

(3.4) $$A_t = b_0 + b_1 E_{pt} E_{\varepsilon t} + b_2 \tilde{E}_{rt},$$

where

$$\tilde{E}_{rt} = \sum_{i=1}^{k} \sum_{j=1}^{k} d_i d_j P_{t-i} \varepsilon_{t-j}$$

$$= E_{pt} E_{\varepsilon t} - E_{rt}^*,$$

where E_{rt}^* follows the same form as E_{rt} except that the geometric parameter is the square of the one considered in E_{rt}.

Thus, Equation (3.4) is equivalent in form with Equation (3.1) if $b_2 = 0$. Therefore, an appropriate nested hypothesis testing procedure is to estimate Equation (3.4) by standard methods as used above and then simply test the hypothesis $b_2 = 0$.

Following this nested hypothesis testing procedure, however, a great deal of collinearity was generally found between the two variables $E_{pt} E_{\varepsilon t}$ and E_{rt} in Equation (3.4) so that no significant results were forthcoming for either b_1 or b_2. Furthermore, the estimated signs of b_1 and b_2 were generally in conflict and thus implausible. The conclusion suggested by this investigation, which was carried out with the California field crop data used by Just (1974), was therefore that the nested hypothesis approach had very little promise, apparently because not all information is used.

That is, the model in (3.4) does not correspond to (3.2) unless $b_1 = b_2$. Thus, the fully specified alternative hypotheses are $H_0 : b_1 = b_2$ and $H_1 : b_2 = 0$. In the nested approach above, the equality of b_1 and b_2

under H_0 is not recognized. When this equality is appropriately considered, the hypotheses are non-nested and hence the standard results for hypothesis testing are inapplicable. For this reason and to use fully all of the information available for testing the hypothesis, a preferable approach is to appeal to the emerging literature on non-nested hypothesis testing, recently reviewed by MacKinnon (1983).

Non-nested Hypothesis Testing

Non-nested hypothesis testing is a relatively new area of development in statistics dating back only to the work of Cox (1961, 1962). Later, Atkinson (1970) introduced an alternative statistic also based on a suggestion by Cox. It was not until the work of Pesaran (1974) that non-nested hypothesis testing began to come in focus in the context of linear models. Pesaran specifically considered application of the Cox approach for testing the alternative sets of hypotheses

$$H_0 : y = X\alpha + w, \quad w \sim N(0, \sigma^2 I),$$

$$H_1 : y = Z\beta + \delta, \quad \delta \sim N(0, \gamma^2 I),$$

where y is an n x 1 vector of observations on the dependent variable, α and β are alternative parameter vectors of respective dimensions p x 1 and q x 1, X and Z are matrices of observations on alternative sets of explanatory variables with respective dimensions n x p and n x q, and w and δ are corresponding n x 1 vectors of disturbances.

Applying the approach of Cox, Pesaran found that the statistic

$$N_0 = (n/2)\log(\hat{\gamma}^2/\phi_0)$$

$$[(\hat{\sigma}^2/\phi_0^2)\hat{\alpha}'X'M_Z M_X M_Z X\hat{\alpha}]^{-0.5}$$

has asymptotically a standard normal distribution under H_0, where X' denotes the transpose of X, and

$$\hat{\alpha} = (X'X)^{-1}X'y,$$

$$\hat{\sigma}^2 = (1/n)(y-X\hat{\alpha})'(y-X\hat{\alpha}),$$

$$\hat{\beta} = (Z'Z)^{-1}Z'y,$$

$$\hat{\gamma}^2 = (1/n)(y-Z\hat{\beta})'(y-Z\hat{\beta}),$$

$$\phi_0 = \hat{\sigma}^2 + (1/n)\hat{\alpha}'X'M_Z X\hat{\alpha},$$

$$M_X = I - P_X,$$

$$P_X = X(X'X)^{-1}X',$$

$$M_Z = I - P_Z,$$

$$P_Z = Z(Z'Z)^{-1}Z.$$

A large positive value of N_0 favors the null hypothesis while a large negative value of N_0 disfavors the null hypothesis.

By reversing the roles of H_0 and H_1 a similar test statistic N_1,

$$N_1 = (n/2)\log(\hat{\sigma}^2/\phi_1)[(\hat{\gamma}^2/\phi_1^2)\hat{\beta}'Z'M_X M_Z M_X Z\hat{\beta}]^{-0.5},$$

is also obtained which has an asymptotic standard normal distribution under H_1 where

$$\phi_1 = \hat{\gamma}^2 + (1/n)\hat{\beta}'Z'M_X Z\hat{\beta}.$$

Thus, an alternative test is also possible.

In a subsequent paper, Dastoor (1978) applied the approach of Atkinson in the context of linear regression to find another test statistic

$$NA_0 = [(n/2)(\hat{\gamma}^2 - \phi_0)/\phi_0 + (0.5/\phi_0)(y-X\hat{\alpha})'P_Z(y-X\hat{\alpha})]$$

$$[(\hat{\sigma}^2/\phi_0^2)\hat{\alpha}'X'M_Z M_X M_Z X\hat{\alpha}]^{-0.5},$$

which also has a standard normal distribution asymptotically. Although some earlier results by Pereira (1977) showed that the Atkinson test may be inconsistent generally, Dastoor found that the Atkinson test is consistent in the context of linear regression. Again, a large positive value of NA_0 favors H_0 while a large negative value of NA_0 disfavors H_0. Again, the roles of hypotheses may be reversed so that the statistic

$$NA_1 = [(n/2)(\hat{\sigma}^2 - \phi_1)/\phi_1 + (0.5/\phi_1)(y-z\hat{\beta})'P_X(y-z\hat{\beta})]$$

$$[(\hat{\sigma}^2/\phi_1^2)\hat{\beta}'Z'M_X M_Z M_X Z\hat{\beta}]^{-0.5}$$

can be used for an alternative test of H_1 where a positive NA_1 favors H_1 and a large negative NA_1 disfavors H_1.

As pointed out by Dastoor, the statistics N_0 and NA_0 may be calculated by performing a series of five regressions:

(i)   regression of y on X to obtain $X\hat{\alpha}$ and $\hat{\sigma}^2$,

(ii)   regression of y on Z to obtain $\hat{\gamma}^2$,

(iii) regression of $X\hat{\alpha}$ on Z to obtain $M_Z X\hat{\alpha}$ (as

       residuals) and $\hat{\alpha}'X'M_Z X\hat{\alpha}$ (as the sum of

       squared residuals),

(iv)   regression of $M_Z X\hat{\alpha}$ on X to obtain $\hat{\alpha}'X'M_Z M_X M_Z X\hat{\alpha}$

       (as the sum of squared residuals), and

(v)   regression of $y - X\hat{\alpha}$ on Z to obtain

       $(y-X\hat{\alpha})'M_Z(y-X\hat{\alpha})$

       (as the sum of squared residuals).

By performing an additional 3 regressions paralleling steps (iii)-(v), test statistics N_1 and NA_1 can also be calculated to test the hypotheses with the roles of the null and alternative reversed. Finally, we note in passing that these results have been extended to the nonlinear regression case by Pesaran and Deaton (1978) so that the above method continues to hold validity when, say, a geometric lag parameter is estimated nonlinearly as in the regression of the previous section.

Using this approach, which is designed to compare hypotheses that are non-nested, we now return to the specific problem of determining whether farmers consider the covariance of prices and yields in making supply response decisions.

A Non-nested Test of Supply Response Specifications

In the context of the above method, consider a null hypothesis corresponding to Equation (3.1) or to $b_1 = b_2$ in Equation (3.4) with an alternative hypothesis corresponding to Equation (3.2) or to $b_2 = 0$ in Equation (3.4). To test the hypothesis comprehensively, we used the annual FAO data on production of the three most important cereal crops--wheat, rice and maize. These crops seemed particularly appropriate in the context of the model specified in Equations (3.1) and (3.2) since they tend to be the dominant crops in most areas of the

world in which they are grown; hence, the exclusion of
price or revenue terms for competing crops does not
necessarily lead to mis-specification problems.   Based
on certain tests for consistency and accuracy, the data
published by FAO were pruned to 44 countries for the
years 1966-1975. A summary of the cases thus considered
is given in Table 3.3. For example, 37 of the 44 countries had consistent and complete time series for hectarage, production and price of wheat. Of these, 15
were developing countries, 16 were industrialized countries, and 6 were centrally planned economies, according
to World Bank classification of 1978. The data were
divided into these groups to determine whether there
are differences in the consideration of covariance under
different degrees of development or under different
economic systems.

   In addition, the data were divided according to
other criteria as explained below in an attempt to
remove possible bias due to mis-specification. Specifically, in the models in Equations (3.1) and (3.2) a
negative estimate of a_1 or b_1 may be reflective of mis-
specification since, normally, higher prices or returns
should lead to crop expansion. Since the equations are
estimated without consideration of competing crops,

TABLE 3.3
Summary of the Country Cases Considered in the
Non-nested Hypothesis Testing

	Wheat	Rice	Maize
Total cases	37	25	33
Developing countries	16	4	9
With positive slope	8	4	4
With negative slope	8	0	5
Industrialized countries	15	16	18
With positive slope	10	10	10
With negative slope	5	6	8
Centrally planned countries	6	5	6
With positive slope	2	4	3
With negative slope	4	1	3
Positive slope	20	18	17
With $R^2 \geq 0.6$	7	6	8
With $R^2 < 0.6$	13	12	9
Negative slope	17	7	16
With $R^2 \geq 0.6$	4	2	1
With $R^2 < 0.6$	13	5	15

negative signs on own price and revenue may thus be due
to spurious correlation with price or revenue series of
competing crops in which case the equations are mis-
specified and statistics are biased. Thus, in view of
these possibilities, the equations with positive estimates for a_1 and b_1 are given special consideration. It
should be noted, however, that the equations are estimated with nominal rather than real prices. Suppose,
for example, that inflation is linear so that nominal
prices are related to real prices by

$$P_t = e + f\tilde{P}_t$$

where $\tilde{P}_t$ represents real price at time t. Then the
model in Equation (3.1) becomes

$$A_t = a_0 + a_1 \sum_{j=1}^{k} d_j (e + f\tilde{P}_{t-j}) E_{\varepsilon t} + w_t,$$

$$= a_0 + a_1 e + a_1 f \left(\sum_{j=1}^{k} d_j \tilde{P}_{t-j} \right) E_{\varepsilon t} + w_t,$$

$$= \tilde{a}_0 + \tilde{a}_1 E_{\tilde{p}t} E_{\varepsilon t} + w_t,$$

where $\tilde{a}_0 = a_0 + \tilde{a}_1 e$, $\tilde{a}_1 = a_1 f$, and $E_{\tilde{p}t}$ represents
expected real price. Thus, a negative coefficient esti-
mate for a_1 may simply reflect a case where real price
is moving in a direction opposite to that of nominal
price--not an uncommon event in countries with high
inflation rates. A similar result also holds for the
model in Equation (3.2). Thus, while cases with positive a_1 and b_1 slopes are given special consideration,
the other cases should not be disregarded.

Second, special consideration is given to equations
with high values of R^2 since correctly specified equations would seemingly be associated with high goodness
of fit. That is, if other important variables are
excluded from the equation by the simple specification
in Equations (3.1) or (3.2), a high value of R^2 would be
impossible except with high spurious correlation between
the included and excluded variables. Furthermore, if
spurious correlation is high, the results of the hypothesis test would tend not to be distorted.

With these considerations in mind, the models in
Equations (3.1) and (3.2) were estimated by a maximum
likelihood search over the parameter space for values of

θ corresponding to 0.1, 0.2, ...,0.9. Although the numbers of individual cases are too numerous to discuss and present individually, the results for the individual cases corresponding to the six largest crop areas are given in Tables 3.4 and 3.5. The detailed statistics in Table 3.4 are intended to permit direct interpretation of some of the supply functions, for the purpose of concrete illustration. Our discussion, however, is concentrated on the types of summary statistics reported in Table 3.5. These six cases account for 82 percent (in terms of 1969 hectarage) of the total crop area involved in the 95 cases considered. The four test statistics, N_0, NA_0, N_1 and NA_1, correspond to the definitions above. As indicated earlier, all four statistics are essentially designed to test the same pair of hypotheses and are, asymptotically, standard normal distributed[4]. As evidenced by the results in Table 3.5, the implications of all four statistics are also essentially the same in every case (the signs N_1 and NA_1 are opposite to the signs of N_0 and NA_0 because the roles of null and alternative hypotheses are reversed). Thus, the discussion of the results in the remainder of this chapter will be confined to N_0, the Cox statistic corresponding to H_0 specified by Equation (3.1) and H_1 specified by Equation (3.2).

From Table 3.5 it may be concluded that results for the six large crop area cases are not significantly in favor of the price-yield model in (3.1) for any of the cases but are significantly in favor of the revenue model in (3.2) at a 2.5 percent level in one case and at a 15 percent level in another. Furthermore, it is also noteworthy that three of the six cases in Table 3.5 represent centrally planned economies where, in fact, no covariance of prices and yields may be transmitted to farmers by the central pricing systems. In two of the three remaining cases, the results are substantially in favor of the revenue model.

One must bear in mind, however, that the statistics in Table 3.5 are based on small samples but yet have only asymptotic properties. To increase the applicability of the asymptotic distribution theory, one can consider averaging the test statistics over groups of cases and then employing the central limit theorem (assuming that the data generating the individual statistics are mutually independent). Using this approach, statistics of the form

$$\sum_{i=1}^{n} e_i / n^{0.5}$$

where e_i represents the N_0 statistic for individual cases within a group, are formed for various groups of

TABLE 3.4
Regression Results for the More Important Countries and Crops
in the Non-Nested Hypothesis Testing

Country	Crop	H_0 (Price and Yield)			H_1 (Revenue)			θ [a]
		a_0	a_1	R^2	b_0	b_1	R^2	
Soviet Union	Wheat	71.47[b] (2.19)	-10.131 (2.456)	0.708 (-0.020)[c]	71.42 (2.17)	-7.894 (1.904)	0.711 (-0.016)	0.6
China, People's Republic of	Rice	28.73 (1.37)	0.637 (0.173)	0.660 (0.015)	28.74 (1.37)	0.636 (0.173)	0.660 (0.015)	0.1
	Wheat	25.22 (0.81)	1.427 (0.315)	0.746 (0.012)	25.22 (0.81)	1.425 (0.314)	0.750 (0.012)	0.1
United States	Maize	20.94 (1.06)	1.334 (0.349)	0.676 (0.018)	20.88 (1.08)	1.274 (0.333)	0.676 (0.017)	0.4
	Wheat	13.77 (0.56)	7.085 (0.457)	0.972 (0.049)	13.68 (0.56)	6.302 (0.402)	0.972 (0.044)	0.5
India	Wheat	13.80 (2.77)	5.585 (3.644)	0.251 (0.314)	13.45 (2.78)	1.651 (0.998)	0.281 (0.104)	0.9

[a] Note that the same maximum likelihood estimate of θ was obtained under both hypotheses in each of these cases.

[b] Units of measurement are in local currency per kilogram for price, tons per hectare for yield and area is measured in 10^6 hectares.

[c] Numbers in parentheses under the respective R^2 values are elasticities with respect to subjective mean price evaluated at the arithmetic sample means.

TABLE 3.5
Summary Statistics for the More Important Countries
and Crops in the Non-Nested Hypothesis Testing

Country	Crop	R^2		Test of H_0 (Price & Yield)		Test of H_1 (Revenue)	
		H_0	H_1	Cox (N_0)	Atkinson (NA_0)	Cox (N_1)	Atkinson (NA_1)
Soviet Union	Wheat	0.71	0.71	-0.33	-0.33	0.25	0.25
China, P.R.	Rice	0.66	0.66	-0.58	-0.58	0.58	0.58
	Wheat	0.75	0.75	0.03	0.03	-0.03	-0.03
United States	Corn	0.68	0.68	0.01	0.01	-0.08	-0.08
	Wheat	0.97	0.97	-1.01	-1.00	0.95	0.96
India	Wheat	0.25	0.28	-2.07	-1.96	1.86	1.96

cases in Table 3.6. Following the previous discussion, the classification is according to types of economies and other manifestations in the results potentially due to mis-specification.

Generally, the results in Table 3.6 tend to favor the revenue model for wheat, are neutral for rice, and are somewhat ambiguous for maize. The results for wheat are almost universally in favor of the revenue model while the maize results favor the price-yield model in free economies and the revenue model in centrally planned economies. Where all cases are considered together or grouped according to economic status, the results for maize are more significant than for other crops while, when possibly mis-specified cases are removed, the results for wheat are more significant.

The results also favor the revenue model when the possibly mis-specified cases are included. Regardless of whether the possibly mis-specified cases are removed or not, the only strong support for the price-yield model comes from the developing country group. Perhaps the following interpretation is appropriate. The cases of maize in developing countries are mostly associated with subsistence agricultures. Perhaps, farmers do not perceive the covariance of prices and yields in such situations because not as large a share of the crop enters the cash market. Indeed, in every comparison in Table 3.6 the results for the developing countries are relatively in favor of the price-yield model which

TABLE 3.6
Standard Normal Cox Test Statistics for H_0
(Price and Yield) in the Non-Nested Hypothesis Testing

	Wheat	Rice	Maize
All Cases			
All Countries	-0.94(37)[a]	0.84(25)	1.36(33)
Developing Countries	-0.42(15)	0.92(16)	2.05(18)
Industrialized Countries	-0.93(16)	-0.25(4)	0.85(9)
Centrally Planned Economies	-0.15(6)	0.57(5)	-1.39(6)
Cases with Positive Slope and $R^2 \geq 0.6$			
All Countries	-2.60(7)	-0.57(6)	0.86(8)
Developing Countries	-1.48(3)	-0.27(1)	1.63(4)
Industrialized Countries	-2.50(3)	-0.48(2)	0.49(2)
Centrally Planned Economies	0.03(1)	-0.37(3)	-1.07(2)

[a] The number of items entering an average is given in parentheses.

excludes covariance, while for the industrialized countries the results are relatively in favor of the revenue model which appropriately reflects covariance of prices and yields. This interpretation of the results also explains to some extent why the results for maize might be more favorable to the price-yield model than for wheat, as well as why the rice results may be somewhere in between. That is, maize tends to be a more important crop in subsistence agricultures, such as the African countries (aside from a few industrialized countries like the United States), while wheat is an important crop mostly in the more developed countries. Thus, if farmers become more aware of the price-yield covariance as commercial agriculture develops, then one would indeed expect the results for wheat to be more in favor of the revenue model, and the results for maize to be more in favor of the price-yield model.

CONCLUSIONS

Although the conventional econometric procedures demonstrated earlier in this chapter are apparently not capable of distinguishing between the alternative price or revenue specifications with typical supply response data, the specific non-nested methods of the latter part of the chapter seem to be more promising. A desirable future research topic would involve extending this analysis to data spanning longer periods of time. Unfortunately, however, reliable longer series of the required types of data exist mostly only for the industrialized countries and thus could not be used to address some of the important questions raised in Table 3.6 regarding response in developing and centrally planned economies.

The results obtained with the present data are not overwhelmingly conclusive, but they are at least consistent with a plausible interpretation of the role of price-yield covariance in farmers' decisions. This interpretation is that farmers in industrialized countries tend to respond to price-yield covariance while subsistence farmers and producers in centrally planned economies tend not to do so. In the former case, there seemingly tends to be a paucity of markets characterized by the distortions discussed by Hazell and Scandizzo as warranting intervention. In the latter case, where this particular type of distortion is implied to occur more widely, there is more promise for correcting distortions. However, even in these cases one must question whether the lack of response to covariance may be due simply to the absence of transmittal of covariances to individual farmers (because many of them participate minimally in commercial markets) or, in the centrally planned economies, because of lack of ability to respond.

NOTES

1. Professor Richard E. Just of the University of California at Berkeley contributed substantially to the research leading to this chapter.

2. Reutlinger (1964), in a minimax criterion framework, used the maximal range of price spread over a specified number of previous years. Freebairn and Rausser (1975) and Lutz (1978) used a moving standard deviation of past prices (i.e. similar to the procedure of Behrman) while Ryan (1977) and Traill (1978) have experimented with a weighted average of absolute deviations of observed prices from past dynamically expected prices.

3. See, for instance, the work of Brownlee and Gainer (1949), Heady (1952), Heady and Kaldor (1954), Kaldor and Heady (1954), Schultz and Brownlee (1942), and Williams (1951).

4. Strictly speaking, one may view the Atkinson statistics with more skepticism than the Cox statistics because the consistency of the Atkinson test has not been shown in the specific regression equation of this study. Consistency was only shown by Dastoor for linear

4
Farmers' Expectations and Mathematical Programming Models of Market Equilibria

Linear programming models can be useful tools for the sector analysis of agricultural supply response and agricultural investment programs. For many purposes it is desirable that such models provide the perfectly competitive solution for all product markets when both prices and quantities are endogenous. Samuelson (1952) provided the basic framework for achieving this in the deterministic case by utilizing the sum of consumers' and producers' surplus as the model maximand. He instigated this result in the context of spatial equilibrium models and Takayama and Judge (1964, 1971) further developed this objective function to obtain quadratic programming formulations for multiproduct models. Duloy and Norton (1973, 1975) subsequently applied the method to agricultural sector models using linear programming approximations.

Hazell and Scandizzo (1974, 1977) have shown how risk-averse behavior can be incorporated into such price-endogenous sector models. However, as has been shown in previous chapters, the nature of the market equilibrium solutions obtained must depend on assumptions about how producers form their expectations about prices and yields. The purpose of this chapter is to provide a consolidated review of the pertinent programming methods and to elaborate the linear programming formulations corresponding to both price and revenue expectations behavior. These formulations are illustrated with a linear programming models of agricultural production at a subsector level in Mexico. The results provide some indication of the social welfare cost of inappropriate price forecasts.

PROGRAMMING MODELS INCLUDING RISK AND VARIOUS
EXPECTATIONS

A Deterministic Formulation

We choose as the basic model into which to incorporate risk and risk-averse behavior of producers a model
in which prices are determined endogenously. In formulating their model, Duloy and Norton (1975) ignored
risk and assumed that farmers maximize profit and com-
pete in a perfectly competitive way. The latter assumption is tantamount to saying that farmers plan on the
basis of constant anticipated prices.

Let an f subscript denote the fth individual farm,
and define p^*_f = an n x 1 vector of anticipated
product prices

c_f = an n x 1 vector of unit costs

x_f = an n x 1 vector of enterprise levels

M_f = an n x n diagonal matrix of enterprise yields with jth diagonal entry
m_{jf}

$q_f = M_f x_f$ is the n x 1 vector of total
outputs.

The objective function for an individual farmer's
planning problem is assumed to be

$$(4.1) \qquad \max_{x_f} u_f = p^{*'}_f q_f - c'_f x_f,$$

(where, in this chapter, a prime denotes a transposed
vector or matrix) subject to a set of linear constraints
denoted by $D_f x_f \leq b_f$. Here D_f is a matrix of unit activity resource requirements, and b_f is a vector of
resource supplies.

The Lagrangian of the problem is

$$(4.2) \qquad L_1 = p^{*'}_f M_f x_f - c'_f x_f + v'_f (b_f - D_f x_f),$$

where v_f is a vector of Lagrange multipliers or dual
values. The necessary Kuhn-Tucker conditions for an
optimal solution can be used to derive the profit maximizing conditions for the individual farm. These necessary conditions are

$$(4.3) \qquad \partial L_1 / \partial x_f \leq 0, \quad \partial L_1 / \partial v_f \geq 0.$$

and

(4.4) $x_f \partial L_1 / \partial x_f = 0, \quad v_f \partial L_1 / \partial v_f = 0.$

The requirements in (4.4) are the complementary slackness requirements that (a) an activity cannot be active and at the same time have a non-zero opportunity cost and (b) a resource cannot be slack and at the same time have a non-zero dual value. Applying the necessary conditions in (4.3) to (4.2) gives

(4.5) $\partial L_1 / \partial x_f = p_f^{*'} M_f - c_f' - v_f' D_f \leq 0$

and

(4.6) $\partial L_1 / \partial v_f = b_f - D_f x_f \geq 0.$

Equation (4.6) is the feasibility requirement and (4.5) is the classical marginality rule for determination of optimal output.

Taking the jth element of the vector $\partial L_1 / \partial x_f$, rearranging terms and dividing by m_{jf}, the jth element of M_f, we obtain

(4.7) $p_{jf}^* \leq (1/m_{jf}) (c_{jf} + \Sigma_r v_{rf} d_{rjf})$

This states that, for each product, the marginal cost per unit of output must be equal to or greater than the anticipated price. The marginal cost comprises own marginal cost c_{jf}/m_{jf} plus resource opportunity costs $(1/m_{jf})\Sigma_r v_{rf} d_{rjf}$ as reflected in the dual values of the resources used by that activity.

While (4.7) is a necessary condition, it is apparent from the complementary slackness conditions, $x_f \partial L_1/\partial x_f = 0$ in (4.4) that Equation (4.7) will always be satisfied as an equality in an optimal solution for the activities which enter the basis, that is, for all $x_{jf} > 0$. Consequently, for all non-zero activities, the marginality rule can be stated as

(4.8) $p_{jf}^* = (1/m_{jf}) (c_{jf} + \Sigma_r v_{rf} d_{rjf}).$

The farm planning problem above is based on a fixed set of anticipated prices where farmers are assumed to be price takers. In a sectoral model in which prices are endogenous it is required to arrive at a set of equilibrium prices and outputs, given downward-sloping demand curves and an aggregate of individual price-taking farmers. Let x, q, M, c and p* be suitable (in the sense elaborated by Day [1963]) aggregates of the individual farm x_f, q_f, M_f, c_f and p_f^* matrices and let the market demand system be denoted by $p = f(q)$, where q is a vector of product demands, and p is a vector of market prices. Then the problem is to find a price vector p which clears all the markets and enables all farmers to meet their profit maximizing conditions in Equation (4.8) with $p = p^*$.

Duloy and Norton solved this problem under two conditions. The first condition is that the demand system is linear of the form, $p = f(q) = a - Bq$, where a is a vector of demand intercepts and B is a symmetric (see Zusman 1969) and positive semi-definite matrix of slope coefficients. The second condition is that farmers' anticipations about prices converge to the market clearing prices.

The model which provides the required solution is

(4.9) $\max\limits_{x} U = q'(a - 0.5Bq) - c'x,$

such that

(4.10) $Dx \leq b$

and

(4.11) $q = x'M,$

where the constraint set (4.10) is understood to denote an appropriate (usually block diagonal) aggregate of the constraint sets for individual farms, and (4.11) contains the market clearing conditions equating the supply and demand of each commodity.

The objective function (4.9) comprises two components. The term $q'(a - 0.5Bq)$ is the sum of areas under the product demand functions and the term $c'x$ is the total cost of production. Since the latter must equal the sum of areas under the product supply functions, the objective function is the sum of producers' and consumers' surplus over all markets.

To show that (4.9), (4.10) and (4.11) indeed yield the required solution, form the Lagrangian function

$$L_2 = x'M(a - 0.5BMx) - c'x + v'(b-Dx)$$

and apply the necessary Kuhn-Tucker condition $\partial L_2/\partial x \leq 0$. This yields

$$Ma - MBMx - c - D'v \leq 0,$$

or, after some rearrangement,

(4.12) $(a - BMx) \leq M^{-1}(c + D'v).$

But a - BMx is the required vector of market clearing prices p and, by the complementary slackness conditions, the jth elements of (4.12) must hold as a strict equality for all nonzero x_j -- that is,

(4.13) $p_j = (1/m_j) (c_j + \Sigma_r v_r d_{rj})$.

If the conditions for aggregation have been met, then from comparison of Equations (4.8) and (4.13), farmers are fulfilling their profit maximizing conditions at the market equilibrium.

Introduction of Risk at the Farm Level of Modeling

As in other chapters, risk is assumed to enter the market only through stochastic yields. The vector of outputs for the fth individual farm now becomes

$$q_f = N_f x_f,$$

where N_f is an n x n diagonal matrix of stochastic yields with jth diagonal element n_{jf} and $E[N_f] = M_f$.

Stochastic yields imply stochastic supply functions and hence stochastic market prices. It is assumed, however, that input costs and the market demand structure are not stochastic and that the farm linear programming constraints are not affected. The latter assumption could be relaxed since several techniques are available to handle stochastic constraints and these techniques do not involve changing the objective function.

It is further assumed that the individual farmers are averse to risk and that their behavior conforms to maximization of a utility function specified in terms of mean and standard deviation of profit. The objective function for the fth farm each year is

(4.14) $\max_{x_f} u_f = E[p_f' q_f] - c_f' x_f - \phi_f V[p_f' q_f]^{0.5}$,

where ϕ_f is a risk aversion coefficient.

To enumerate (4.14) more fully we should introduce some assumptions about the nature of farmers' perception of the parameters of the joint price and yield distribution, that is, of $E[p_f q_f] = E[p_f N_f] x_f$ and $V[p_f q_f] = x_f' V[p_f N_f] x_f$. Earlier we have considered two particular ways of formulating expectations about $E[p_f' N_f]$, namely, independent price p_f^* and yield N_f^* forecasts, so that $E[p_f' N_f] = p_f^{*'} N_f^*$, and rational expectations based on revenues with $E[p_f' N_f] = r_f^{*'}$, where r_f^* is a vector of anticipated revenues. We showed that these two types of expectations lead to equilibria different in terms of the asymptotic values of expected market clearing prices and outputs. Consequently, different model specifications are required to simulate the market equilibria corresponding to these two behavioral modes. It is also evident that there are doubtless alternative ways of subjectively formulating $V[p'q_f]$ and that these may lead to convergence to different equilibrium values. However, to simplify the analysis, we assume that these expectations are always formed in the same way and that they converge to statistically observable relationships at market equilibrium. Specifically, we assume that $x'_f V[p_f N_f] x_f = x'_f \Omega_f x_f$ where Ω_f is a given covariance matrix of unit revenues for the activities.

Substituting price or revenue expectations for $E[p'_f N_f]$ in Equation (4.14), setting $V[p'_f q_f] = x'_f \Omega_f x_f$, and pursuing algebra analogous to that for the deterministic case, it is straightforward to derive the optimal conditions for the individual farmer. For simple price anticipations, the jth product ($x_{jf} \geq 0$) will be produced until

$$(4.15) \qquad p^*_{jf} n^*_{jf} = c_{jf} + \Sigma_r v_{rf} d_{rjf} + \phi_f (x'_f \Omega_f x_f)^{-0.5}$$
$$\Sigma_i \omega_{jif} x_{if},$$

whereas, for revenue expectations behavior, the corresponding rule is

$$(4.16) \qquad r^*_{jf} = c_{jf} + \Sigma_r v_{rf} d_{rjf} + \phi_f (x'_f \Omega_f x_f)^{-0.5}$$
$$\Sigma_i \omega_{jif} x_{if}.$$

The right-hand sides of Equations (4.15) and (4.16) are identifical and differ from the deterministic counterpart in Equation (4.7) by the addition of a marginal risk term, $\phi_f (x'_f \Omega_f x_f)^{-0.5} \Sigma_i \omega_{jif} x_{if}$. This risk term is another cost, namely, the additional expected return demanded by farmers as compensation for taking risk.

This result is consistent with the results obtained for single-product firms (for example, Magnússon 1969). The appearance of the marginal risk cost provides some rationale for the claim that deterministic models tend to overestimate the supply response of high-risk crops. For such crops, the term $\Sigma_i \omega_{jif} x_{if}$ will tend to be positive and the risk-inclusive marginal cost curve will lie above the curve that would be obtained from a deterministic or risk-neutral model.

It is apparent from the similarity of Equations (4.15) and (4.16) that differences in supply response for such individual farms can only arise from differences between $p^*_{jf} n^*_{jf}$ and r^*_{jf}. Since $r^*_{jf} = E[p_{jf} n_{jf}]$ and p^*_{jf} are subjective means, then the two terms differ by the covariance between price and yield, $cov[p_{jf}, n_{jf}]$. If this covariance is negative (positive), then revenue expectations will lead to a lower (higher) output than when price expectations are held.

We turn now to the problem of formulating price-endogenous models which can simulate the market equilibria corresponding to price and revenue expectations behavior.

Introduction of Risk at the Sectoral Level of Modeling

The matrices x, q, M, c and p^* have already been defined as aggregates of farm-level x_f, q_f, M_f, c_f and p^*_j matrices. Also, let N and N^* denote suitable aggregates of farm-level N_f and N^*_f matrices, with $E[N] = M$, let Γ be a suitable aggregate of farm-level Ω_f matrices, and let Φ be an aggregate of the farm-level risk parameters ϕ_f.

In forming these aggregates, there are definite problems that are closely related to the problem of establishing farm classification criteria for exact aggregation in quadratic models. This problem lies beyond the scope of our present treatment and it is merely noted that the aggregate variables Γ and Φ must be chosen so that

$$\Phi (x' \Gamma x)^{0.5} = \Sigma_f \phi_f (x'_f \Omega_f x_f)^{0.5}.$$

This equation states that the aggregate level of risk adjustment calculated for the model must be equal to the sum of individual risk adjustments over all farms. Without this condition the covariance relationships between farms could be exploited in the aggregate model in seeking efficient diversification and this would be inconsistent with the competitive behavior assumed. For example, if there are k farms which are identical, a suitable choice of the aggregate variables would be $\Phi = \phi_f$ and $\Gamma = \Omega_f$ so that

$$\Sigma_f \phi_f (x'_f \Omega_f x_f)^{0.5} = k \phi_f (x'_f \Gamma_f x_f)^{0.5}$$

$$= \Phi (k x'_f \Gamma k x_f)^{0.5}$$

$$= \Phi (x' \Gamma x)^{0.5},$$

where
$$x = k x_f$$
$$= \Sigma_f x_f.$$

60

Consider now the modified objective function

(4.17)
$$\max_{x} U = x'N^*(a - 0.5BN^*x) - c'x$$
$$- \Phi(x'\Gamma x)^{0.5},$$

where $x'N^*(a - 0.5BN^*x)$ is the sum of areas under the product demand curves given anticipated yields N^* and anticipated supplies N^*x, and $c'x + \Phi(x'\Gamma x)^{0.5}$ is the sum of total costs. In terms of the theory in Chapter 2, $c'x + \Phi(x'\Gamma x)^{0.5}$ is the sum of areas under the anticipated supply functions, and $x'N^*(a - 0.5BN^*x)$ is the sum of areas under the demand curves when evaluated at anticipated supplies N^*x. In fact, (4.17) measures the ex ante social welfare and has its maximum at the point where the demand and anticipated supply schedules intersect.

To verify this, form the Lagrangian function

$$L_3 = x'N^*(a - 0.5BN^*x) - c'x - \Phi(x'\Gamma x)^{0.5}$$
$$+ v'(b-Dx)$$

where b and D denote the aggregate constraints and v is a vector of dual values. Apart from complementary slackness requirements, the necessary Kuhn-Tucker conditions are

(4.18)
$$\partial L_3/\partial x = N^*a - N^*BN^*x - c - \Phi\Gamma x(x'\Gamma x)^{-0.5}$$
$$- D'v \leq 0$$

and

(4.19)
$$\partial L_3/\partial v = b - Dx \geq 0,$$

where (4.19) is the feasibility requirement, and (4.18) can be rearranged as

(4.20)
$$(a - BN^*x) \leq N^{*-1}[c + \Phi\Gamma x(x'\Gamma x)^{-0.5} + D'v].$$

Now N^*x is the vector of anticipated supplies, and $a - BN^*x$ is the corresponding vector of anticipated prices. Furthermore, the right-hand side of (4.20) is the sum of expected marginal cost curves over all farms (the jth component is the sum of right-hand sides of Equations like (4.15)). That is, it is the vector of aggregate anticipated supply functions. The inequality in (4.20) states that, in aggregate, farmers must operate around points on the anticipated supply functions which lie at or above the intersections with demand. By the complementary slackness conditions, optimality occurs at the intersection point for all nonzero activities in the solution, and then $a - BN^*x$ is the intersection price vector.

The relevance of the objective function (4.17) is immediate. Suppose farmers anticipate the mean yields $N^* = M$ and suppose, further, that their price forecasts are based on an average of past prices, so that, at market equilibrium, $p^* = E[p] = a - BMx$. Then, substituting $N^* = E[N] = M$ into (4.17), the maximand

$$(4.21) \qquad \max_{x} U = x'M(a - 0.5BMx) - c'x$$
$$- \Phi(x'\Gamma\ x)^{0.5},$$

provides the competitive market solution in which expected marginal costs are equated to the product of the expected prices and expected yields. This formulation corresponds to price expectations behavior.

We turn now to the formulation of the counterpart model which yields the competitive market equilibrium for revenue expectations.

In Chapter 2, it was shown that maximization of an ex post measure of the sum of producers' and consumers' surplus leads to revenue expectations. Not surprisingly, this ex post measure of welfare is also the relevant maximand for obtaining the competitive market solution in the linear programming context when farmers act on the basis of revenue expectations.

The relevant maximand is

$$(4.22) \qquad \max_{x} E[W] = E[x'N(a - 0.5BNx)] - c'x$$
$$- \Phi(x'\Gamma\ x)^{0.5}.$$

Comparing Equations (4.21) and (4.22), it is evident that the two maximands differ only by the term

$$(4.23) \qquad 0.5\ x'V[NBN]x = 0.5\ \Sigma_i\ \Sigma_j\ x_i\ x_j\ \sigma_{ij}\ b_{ij},$$

where $V[NBN] = E[NBN] - MBM$ is a variance-covariance matrix of weighted yields, with ijth element $\sigma_{ij}\ b_{ij}$, and σ_{ij} denotes the covariances of yields between crops i and j, and b_{ij} denotes the cross-demand coefficient. Given that Equation (4.22) is accepted as a useful welfare function, (4.23) is a direct measure of the social costs of market inefficiency that arises when farmers plan on the basis of price rather than revenue expectations.

We now show that (4.22) provides the solution corresponding to the competitive market equilibrium that would be attained if farmers planned on the basis of revenue expectations. The Lagrangian for the problem is

$$(4.24) \qquad L_4 = x'Ma - 0.5x'E[NBN]x - c'x$$

$$- \Phi(x'\Gamma x)^{0.5} + v'(b - Dx),$$

where v is a vector of dual values. Apart from the feasibility conditions $Dx \leq b$, the necessary Kuhn-Tucker conditions imply that

$$(4.25) \qquad Ma - E[NBN]x \leq c + \Phi\Gamma x(x'\Gamma x)^{0.5} + D'v.$$

From the assumed demand structure, the vector of market clearing prices in any one year is $p = a - BNx$. Multiplying by the diagonal matrix of yields N, the vector of unit revenues r is then $r = Np = Na - NBNx$. Now, providing that farmers form their anticipations about r^* in such a way that, in equilibrium $r^* = E[r] = Ma - E[NBN]x$ (as would happen if they took a weighted average of past revenues), then the left-hand side of (4.25) can be written as $E[r]$, and thus at optimality, expected unit revenue must be equal to or less than marginal cost for each activity. Complementary slackness conditions require that Equation (4.25) hold as a strict equality for all nonzero x_j, in which case, expected marginal revenue equals marginal cost at optimality. Assuming aggregation in the model is exact, then this condition is identical to that derived for the farm problem in (4.16).

Linear Programming Approximations for the Nonlinear Planning Problems

The aggregate models with the objective functions defined in (4.21) and (4.22) are quadratic programming problems. Because of the large dimensions of any realistic sector model and the difficulties that still exist with quadratic programming computer codes in solving large problems, it is desirable to linearize these models.

Both objective functions contain the risk term $\Phi(x'\Gamma x)^{0.5}$. Since this term is to be minimized in (4.21) and (4.22), it can be linearized through separable programming (Thomas et al. 1972). However, the approach leads to large programming problems and must be solved with special separable linear programming algorithms. Alternatively, if Γ is estimated on the basis of time series data, an efficient way of linearizing the quadratic term $\Phi(x'\Gamma x)^{0.5}$ is to use the mean absolute deviation (mad) method proposed by Hazell (1971a).

Let $r_{jt} = p_{jt} n_{jt}$ denote the tth year's observation on the revenue of the jth activity x_j, $t = 1, 2, \ldots, T$, and let $\bar{r}_j$ denote the sample mean revenue for the activity over the T years.[1] Then, the mad estimator of the standard deviation of income[2] is

(4.26) $\qquad \Delta(1/T)\Sigma_t|\Sigma_j (r_{jt} - \bar{r}_j)x_j|$,

where $\Delta = (0.5 \ \pi T/(T-1))^{0.5}$ is Fisher's (1920) correction factor to convert the sample mad to an estimate of the population standard deviation,[3] and π is the mathematical constant.

To obtain a linear programming formulation, define variables $z_t \geq 0$ for all t such that

(4.27) $\qquad z_t + \Sigma_j (r_{jt} - \bar{r}_j)x_j \geq 0$,

where $\Sigma_j(r_{jt}-\bar{r}_j)x_j$ measures the deviation in total revenue from the mean, $\Sigma_j \bar{r}_j x_j$, for the tth set of revenue outcomes. If the z_t variables are selected in a minimizing way for each t, either $z_t = 0$ when $\Sigma_j(r_{jt} - \bar{r}_j)x_j$ is positive, or z_t measures the absolute value of the negative deviation in total revenue when $\Sigma_j (r_{jt} - \bar{r}_j)x_j$ is negative. Consequently, $\Sigma_t z_t$ measures the sum of the absolute values of the negative deviations. Since the sum of the negative deviations around the mean is always equal to the sum of the positive deviations for any random variable, it follows that $2\Sigma_t z_t$ is the sum of absolute deviations in total revenue and hence the mad estimate of $(x'\Gamma x)^{0.5}$ is

(4.28) $\qquad (2\Delta/T)\Sigma_t z_t$.

In summary, the appropriate linear programming subproblem to minimize $\Phi(x'\Gamma x)^{0.5}$ is

(4.29) $\qquad \min K\Sigma_t z_t$,

such that $\qquad z_t + \Sigma_j(r_{jt} - \bar{r}_j)x_j \geq 0$, all t,

where $K = \Phi 2\Delta/T$ is a constant.

The reliability of this method compared to using quadratic programming directly on $\Phi(x'\Gamma x)^{0.5}$ has been discussed elsewhere (Hazell 1971a, Thomson and Hazell, 1972). Basically it depends on the relative efficiency of the sample mad compared to the sample standard deviation as an estimator of the population standard deviation.[4] Surprisingly, the sample mad may actually be better than the sample standard deviation for skewed income distributions (Hazell 1971b), but is less efficient for normal distributions (Hazell 1971a).

The objective function for the price expectations model also contains the quadratic term x'M (a - 0.5BMx). This can be written as $\bar{q}'(a - 0.5B\bar{q})$ where $\bar{q} = E[q] = Mx$ is the vector of mean quantities. Duloy and Norton (1975) have shown how this term can be efficiently linearized.

To illustrate their method, consider the simplest case in which B is diagonal, implying that the product demands are independent.[5] Then, letting V_j denote the area under the demand curve (the definite integral) from 0 to $\bar{q}_j$ for the jth product,

$$\bar{q}' (a - 0.5B\bar{q}) = \Sigma_j (a_j\bar{q}_j - b_j\bar{q}_j^2)$$

$$= \Sigma_j V_j.$$

Now V_j is a quadratic, concave function with respect to $\bar{q}_j$ and, since the programming model is a maximization problem in $\bar{q}'(a - 0.5 B\bar{q})$, then V_j can be approximated by a series of linear segments using conventional linear programming codes.

Duloy and Norton define a discrete number of intervals on the $\bar{q}_j$ axis and assign specific values of $\bar{q}_j$, say $\bar{q}_{ij}$, as the boundary points for the intervals. To each of these interval boundary points is then assigned a value of $V_j = a_j \bar{q}_{ij} - b_j \bar{q}_{ij}^2$, which will be designated by v_{ij}. Then, by defining new segment-weighting activities, V_{ij}, $0 \leq V_{ij} \leq 1$, $i = 1, 2,..., k$, the part of the programming problem involving max $\bar{q}'(a - 0.5 B\bar{q})$ is replaced by the linear programming problem

(4.30) $$\max \Sigma_j \overset{k}{\underset{i=1}{\Sigma}} v_{ij} V_{ij},$$

such that

(4.31) $$x_j m_j - \Sigma_i \bar{q}_{ij} V_{ij} \geq 0, \text{ all } j,$$

and

(4.32) $$\Sigma_i V_{ij} \leq 1 \qquad , \text{ all } j.$$

This method adds only two rows for each product but permits inclusion of as many V_{ij} activities as necessary to increase the accuracy of the approximation to any desired degree.

The objective function for the revenue expectations model contains the quadratic term $E[q'(a - 0.5Bq)]$, where $q = Nx$ is the vector of realized quantities. It has already been seen that this objective function differs from the corresponding objective function for price expectations only by the quadratic term $0.5x'V[NBN]x$. One approach to linearizing $E[q'(a - 0.5Bq)]$ is therefore to use separable linear programming techniques to linearize $0.5x' V [NBN]x$ and to subtract this term from the linearized formulation of (4.21). However, when B is diagonal, Hazell and Pomareda (1981) have shown that a simpler approach can be taken.

Since $q = x_j n_j$, then $E[q_j] = \bar{q}_j = x_j m_j$ and $x_j^2 = \bar{q}_j^2/m_j^2$. Further, since $q_j^2 = x_j^2 n_j^2$, then $E[q_j^2] = x_j^2 E[n_j^2] = \bar{q}_j^2 (E[n_j^2]/m_j^2) = \bar{q}_j^2 (1 + C[n_j]^2)$, where $C[n_j]$ is the coefficient of variation of yield. That is, $C[n_j]^2 = (V[n_j]/m_j^2) = (E[n_j^2] - m_j^2)/m_j^2$.

Now if B is diagonal,

$$E[q'(a - 0.5Bq)] = \bar{q}'a - 0.5 E[q'Bq),$$

$$= \bar{q}'a - 0.5 \sum_j b_j E[q_j^2]$$

$$= \bar{q}'a - 0.5 \sum_j b_j \bar{q}_j (C[n_j]^2 + 1)$$

$$= \bar{q}' (a - 0.5BC\bar{q})$$

where C is a diagonal matrix with jth diagonal element $C[n_j]^2 + 1$. Since C is a matrix of constants then BC can be calculated as part of the input to the model. The matrix BC remains diagonal and the term $\bar{q}'(a - 0.5BC\bar{q})$ can be linearized using the Duloy and Norton method.

To summarize, if B is diagonal the linearized models can both be portrayed as:

(4.33)

$$\max U = \sum_{j=1}^{n} \sum_{i=1}^{k} v_{ij} V_{ij} - \sum_{j=1}^{n} c_j x_j$$

$$- K \sum_{t=1}^{T} z_t,$$

such that

(4.34)

$$x_j m_j - \Sigma_j \bar{q}_{ij} V_{ij} \geq 0, \text{ all } j,$$

(4.35)

$$\Sigma_i V_{ij} \leq 1, \text{ all } j,$$

(4.36)

$$z_t + \Sigma_j (r_{jt} - \bar{r}_j)x_j \geq 0, \text{ all } t,$$

and

(4.37)

$$Dx \leq b$$

where V_{ij}, x_j and z_t are the unknown activity levels.

The difference between the two models is that $v_{ij} = a_j \bar{q}_{ij} - b_j \bar{q}^2_{ij}$ when price expectations are specified, but $v_{ij} = a_j \bar{q}_{ij} - b_j (C[n_j]^2 + 1)\bar{q}^2_{ij}$ when revenue expectations are specified.

Table 4.1 shows the layout of the linear programming tableau corresponding to Equations (4.33) to (4.37).

TABLE 4.1
Layout of Tableau for the Linearized Models

Constraint and Text Equation Numbers	Production Activities x_1 x_2 ... x_n	Activities to Linearize Areas Under Demand $v_{11}...v_{1k}...v_{n1}...v_{nk}$	Negative Deviation Counters for Revenue $z_1 ... z_T$	RHS
Objective Function (4.33)	$-c_1$ $-c_2$ ... $-c_n$	$v_{11}...v_{1k}...v_{n1}...v_{nk}$	$-K ... -K$	Maximize
Resource Constraints (4.37)	D Matrix			$\leq$ b Vector
Commodity Balance Constraints (4.34)				
Crop $j = 1$	m_1	$-q_{11}...-q_{1k}$		≥ 0
$\vdots$	$\ddots$	$\ddots$		$\vdots$
$j = n$	m_n	$-q_{n1}...-q_{nk}$		≥ 0
Convex Combination Constraints (4.35)				
Crop $j = 1$		$1 ... 1$		≤ 1
$\vdots$		$\ddots$		$\vdots$
$j = n$		$1 ... 1$		≤ 1
Revenue Deviation Constraints (4.36)				
Year $t = 1$	$r_{11}-\bar{r}_1$ $r_{21}-\bar{r}_2...r_{n1}-\bar{r}_n$		1	≥ 0
$\vdots$	$\vdots$		$\ddots$	$\vdots$
$t = T$	$r_{1T}-\bar{r}_1$ $r_{2T}-\bar{r}_2...r_{nT}-\bar{r}_n$		1	≥ 0

AN ILLUSTRATIVE APPLICATION IN MEXICO

The Model

An agricultural sector model, CHAC, already existed for Mexico (Duloy and Norton 1973) and provided a suitable basis for comparing the equilibrium effects of price versus revenue expectations behavior. CHAC is a linear programming model which encompasses the supply--domestic and imported--and all demands--domestic and export--for 33 short-cycle crops. It does not include livestock, forestry or long-cycle crops. The model is an aggregate of regional submodels that are linked through a national market structure (domestic and foreign) and by some common resource constraints. CHAC is a static equilibrium model and provides the perfectly competitive solution to all markets for both prices and quantities through use of the kind of maximand detailed in Equation (4.9) under the assumption that farmers are, on the average, risk neutral ($\Phi = 0$).

To keep this illustrative study within manageable limits, a small version of CHAC was created which included only selected areas of irrigated land. These selected areas represent eight of the more than one hundred administrative districts of the Mexican Ministry of Water Resources. They are not contiguous districts but are scattered throughout the arid agricultural areas of Mexico. The districts are, by regions: Pacific Northwest -- Culiacán, Comisión del Fuerte, Guasave, Rio Mayo, Santo Domingo; North Central -- Ciudad Delicias, La Laguna; Northeast -- Bajo Rio San Juan. Taken together, these districts account for significant shares of the national production of cotton, tomatoes, dry alfalfa, rice, soybeans and safflower (see Table 4.2). The average district cropping patterns for the years 1967/68 to 1969/70 are given in Table 4.2, but exclude a small percentage of land devoted to crops that are not included in the models. Crop production is almost entirely dependent on irrigation in all eight districts, and any small areas of rainfed land have been excluded.

In total, the eight district models cover 99,000 farms of an average size of 5.8 hectares--a district breakdown is included in Table 4.2. For modeling purposes, each district is treated as a single large farm. The farms are thought to be sufficiently homogenous that this procedure is unlikely to lead to any serious aggregation bias problems. The model activities provide for the production, in each district, of the crops grown by that district in Table 4.2, each with a choice of three mechanization levels and two planting dates. A set of labor activities provide flexibility in selecting seasonal combinations of family and hired day labor.

TABLE 4.2
Average District Cropping Patterns, 1967/68 to 1969/70
(Harvested Hectares)

Crops	El Fuerte	Culiacán	Río Mayo	Guasave	Delicias	San Juan	St. Domingo	Laguna	Aggregate	Proportion of National Production (Percent)
Dry Alfalfa	1,988	–	2,144	–	6,510	–	285	5,498	16,425	34
Cotton	46,364	–	15,535	–	7,903	1,190	17,585	67,964	156,541	25
Green Alfalfa	–	543	–	3,480	–	–	–	5,224	5,767	2
Rice	11,335	23,568	–	3,480	–	–	–	–	38,383	25
Sugar cane	12,706	24,172	–	–	–	–	–	–	36,878	12
Safflower	4,790	13,374	10,435	3,737	–	–	1,098	–	33,434	29
Barley	–	–	112	–	–	–	–	–	112	1
Chilies	386	1,570	–	48	–	–	–	–	2,004	9
Beans	16,224	11,024	–	202	–	–	–	–	27,450	3
Chickpeas	561	938	–	271	–	–	–	–	1,770	1
Tomatoes	3,049	9,563	–	581	–	–	–	–	13,193	37
Sesame	3,010	2,815	8,390	144	–	–	–	–	14,359	4
Maize	10,792	4,302	4,071	2,420	10,053	54,269	1,038	6,213	93,158	2
Cantaloupe	231	397	–	722	–	–	–	–	1,340	4
Potatoes	1,320	–	–	–	–	–	–	–	1,320	5
Cucumbers	–	–	–	8	–	–	–	–	8	0
Watermelons	757	325	–	41	–	74	–	–	1,197	5
Sorghum	24,238	22,795	10,616	1,238	7,719	19,876	–	5,592	92,074	11
Soybeans	16,264	4,392	11,886	–	–	–	–	–	32,542	20
Wheat	23,561	3,057	29,969	5,742	29,668	1,048	11,738	16,150	120,933	16
TOTALS	177,576	122,825	93,158	18,634	61,853	76,457	31,744	106,641	688,888	
Number of Farms	16,484	6,224	9,185	2,984	10,710	4,480	647	48,341	99,055	
Available Hectares Per Farm	10	12	8	6	4	16	47	2	n.a.[a]	

[a] n.a. means not available.

Family labor is charged a reservation wage of one-half of the hired day labor rate. Purchasing activities provide for the supplies of mules, machinery and irrigation water. Seasonal constraints are imposed on land and labor, and an annual constraint is imposed on water supplies. Technical coefficients and costs are taken at 1967/68 to 1969/70 average levels. The model constraints are also based on this period. Average yields are based on the six-year period 1966/67 to 1971/72, and risk parameters were estimated from time series data spanning the period 1961/62 to 1970/71.

The district models are linked in block diagonal form and integrated into an aggregate market structure, similar to that in CHAC. That is, the market comprises linear domestic demand functions[6] of the form p = a - BNx, and has import and export possibilities at fixed prices. The domestic demand curves have the same national price elasticities as in CHAC (see Table 4.3), but are located at mean output levels appropriate for the eight district aggregates. Export and import constraints are also pro-rated according to the ratio of output from the eight districts to national output for each product. Finally, the model structure was changed from that of CHAC by introducing risk-averse behavior. Specifically, farmers are assumed to maximize utility described by the function $E(y) - \Phi S(y)$.

The resultant model was solved for competitive equilibrium solutions to all product markets corresponding to price, and then revenue expectations behavior. This was achieved by using the two linear programming formulations described in the previous section. In both cases the aggregate risk aversion parameter Φ was varied in order to evaluate the effects of different levels of risk-averse behavior on the model solutions.

The Results

Tables 4.3 and 4.4 show the domestic price solutions to the two models for different values of Φ. These values are the expected market clearing prices in equilibrium, and are obtained as dual values from the model solutions. Since there is obviously a one-to-one relationship between the market prices and the total production of each crop sold in the domestic market, differences between price solutions also reflect differences in production.

In general, the revenue expectations model leads to higher domestic prices, and hence to lower production for the domestic market. This effect is particularly pronounced in the case of chilies, while wheat and maize provide the only significant exceptions. This result is to be expected, since revenue expectations for risky

TABLE 4.3
Price Solutions for the Price Expectations Model with Various Φ Values
(Pesos/Tonne)

Demand Group	Commodity	Values of Φ					Actual Base Period Prices	Group Own Price Elasticity
		0	0.5	1.0	1.5	2.0		
1.	Sugar Cane	68	68	70	68	69	70	-0.25
2.	Tomatoes	330	705	1,071	1,319	1,636	1,150	-0.4
3.	Chilies	700	741	748	828	965	1,500	-0.2
4.	Cotton Fiber	5,770a	5,770a	5,770a	5,770a	5,770a	5,770a	-0.5
5.	Dry Alfalfa	308	347	442	485	498	400	
	Green Alfalfa	75	69	64	63	62	100	
	Barley	848	787	793	806	801	930	-0.3
	Chickpeas	1,725	1,673	1,631	1,818	1,987	990	
	Maize	1,044	981	1,023	1,106	1,128	860	
	Sorghum	557	568	549	533	549	630	
6.	Rice	1,195	1,101	1,047	1,035	1,038	1,220	
	Beans	2,334	2,179	2,091	2,053	2,040	1,830	-0.3
	Chickpeas	1,725	1,673	1,631	1,818	1,987	990	
	Potatoes	391	468	537	699	882	930	
7.	Maize	1,044	981	1,023	1,106	1,128	860	-0.1
	Wheat	971	912	905	935	959	800	
8.	Cantaloupe	329	488	705	935	928	680	-2.0
	Watermelons	297	291	260	299	280	780	
9.	Safflower	1,550	1,401	1,391	1,484	1,509	1,550	-1.2
	Sesame	3,444	3,384	3,151	3,000	2,989	2,410	
	Cotton Oil	651	690	796	1,047	1,206	830	
	Soybeans	1,382	1,364	1,356	1,422	1,466	1,600	
10.	Cucumbers	569	681	838	790	317	590	-0.6
Social Welfare (Billions of Pesos)		4.21	4.04	3.97	3.83	3.83	n.a.b	

a Export price.
b n.a. means not available.

TABLE 4.4

Price Solutions for the Revenue Expectations Model with Various Φ Values
(Pesos/Tonne)

Demand Group	Commodity	Values of Φ					Actual Base Period Prices
		0	0.5	1.0	1.5	2.0	
1.	Sugar Cane	75	73	74	73	74	70
2.	Tomatoes	428	756	1,150	1,387	1,729	1,150
3.	Chilies	2,927	2,980	3,071	3,233	3,380	1,500
4.	Cotton Fiber	5,770a	5,770a	5,770a	5,770a	5,770a	5,770a
5.	Dry Alfalfa	492	465	467	459	472	400
	Green Alfalfa	139	103	90	88	90	100
	Barley	860	791	776	742	761	930
	Chickpeas	1,725	1,595	1,535	1,588	1,720	990
	Maize	1,011	957	1,016	1,000	1,036	860
	Sorghum	605	546	505	515	542	630
6.	Rice	1,424	1,377	1,322	1,247	1,256	1,220
	Beans	2,259	2,082	2,023	1,929	1,992	1,830
	Chickpeas	2,063	2,106	2,068	2,105	2,218	990
	Potatoes	1,900	2,341	2,241	2,216	2,296	930
7.	Maize	1,027	1,011	1,045	1,079	1,132	860
	Wheat	955	899	899	896	909	800
8.	Cantaloupe	528	749	454	743	734	680
	Watermelons	521	499	520	490	463	780
9.	Safflower	1,567	1,494	1,489	1,449	1,451	1,550
	Sesame	3,548	3,360	3,043	3,048	3,093	2,410
	Cotton Oil	747	761	818	1,003	1,171	830
	Soybeans	1,495	1,466	1,437	1,477	1,467	1,600
10.	Cucumbers	579	702	291	108	6	590
Social Welfare (Billions of Pesos)		4.48	4.27	4.10	3.96	3.85	n.a.b

a Export price.
b n.a. means not available.

crops are lower than price expectations because of the
negative covariances between prices and yields. Revenue
expectations therefore lead to lower supplies and higher
market clearance prices on average. The perverse outcomes for wheat and maize demonstrate ambiguities that
arise in the multiproduct case from the existence of
cross-demand effects and negative correlations among
yields.

The bottom rows of Tables 4.3 and 4.4 show the value
of social welfare obtained from the two models for different values of Φ. Welfare is measured using Equation
(4.22), that is, using the expected value of the sum of
'realized', or ex post producers' and consumers' surplus. This value was obtained directly from the objective function for the revenue-expectations model but had
to be calculated from the solution to the price expectations model.

These results demonstrate the superiority of market
equilibria associated with revenue expectations behavior. For $\Phi = 0$, for example, welfare is 270 million
pesos greater when farmers plan on the basis of revenue
rather than price expectations.

The difference in welfare between the two sets of
solutions diminishes as Φ increases, suggesting that
private risk-averse behavior on the part of farmers can
offset, to some extent, the shortcomings of price-expec-
tations behavior. Indeed, when $\Phi = 2.0$, the two models
have almost identical welfare outcomes.

Various simple measures of the overall goodness of
fit between the different price solutions and the base-
period actuals were used, including the mean absolute
deviation, and the correlation and the regression coefficients between the model predictions and the base-year
prices. To illustrate, the goodness of fit in prices
for the two models is shown below as measured by the
mean absolute deviation in model prices from base year
actuals.

Φ	Price Model	Revenue Model
	(mad in pesos/tonne)	
0.0	330	365
0.5	292	351
1.0	257	329
1.5	278	354
2.0	303	413

These mads, as well as the other indicators used,
show that the price expectations model provides a superior fit for all values of Φ. Furthermore, both models
provide their best fit when $\Phi = 1.0$, a plausible value
given the underlying $E[y] - \Phi S[y]$ utility specification.

We do not pretend that this is an ideal method of ascertaining the nature of farmers' expectations. Misspecification in the model, and/or incorrect data, could lead to preferred results for the wrong reasons. Further, the simple mad measure of goodness of fit used here is open to criticism in that we have only taken account of product price deviations (we ignore the model's fit in quantities, trade and factor prices, for example) and each price deviation has been given equal weight irrespective of the corresponding base-year price. Nevertheless, the difference in the goodness of fit between the two models is substantial, and the superiority of the price model is consistent with the econometric results for developing countries reported in Chapter 3.

CONCLUSION

In this chapter we have been concerned with the formulation of mathematical programming models which provide the competitive equilibrium values of expected prices and quantities in multiple product markets when yields are risky. Such models are useful for agricultural sector analysis where it is necessary to take account of the effects of aggregate supplies on market prices.

Our analysis was simplified by assuming that the market demand structure is linear, and that farmers maximize a mean-standard deviation utility function in income. Models were formulated corresponding to underlying price and revenue expectations, and differences between the two models were discussed. Since the two models turned out to be quadratic programming problems, we also discussed ways in which they could be approximated by linear programming.

Finally, we applied both the price and revenue models at an aggregate, but subsector, level in Mexico. Differences between the model results were quite substantial, with the revenue model generally leading to higher expected market prices, lower expected supplies and higher values of social welfare. However, the price model seemed to describe actual market prices somewhat better, supporting the econometric evidence in Chapter 3 that farmers in less developed countries tend to ignore covariances between prices and yields.

NOTES

1. The raw data should first be analyzed for any systematic trends over time and these components removed to obtain a random residual.

2. Since we have assumed that costs are non-stochastic, the variances of income and total revenue are identical.

3. Strictly speaking, this form of Fisher's correction factor only holds when total revenue is normally distributed. For other distributions, appropriate correction factors should be used (see Fisher 1920).

4. More favorable results about the ability of the sample mad to rank farm plans efficiently (Thomson and Hazell 1972) do not hold in the current problem because the mad estimate of $(x'\Gamma x)^{0.5}$ is required to appear in objective functions in which its numerical value is important.

5. Full B matrices can also be linearized provided that they are symmetric. Alternatively, a rather more crude but effective procedure is to group commodities into demand independent groups and to treat each group as a single commodity. See Duloy and Norton (1975) for a discussion of these methods.

6. To approximate cross-elasticity relationships in demand, the crops are classified into demand independent groups (see Table 4.3), and linear substitution is allowed between products within each group at rates fixed by average 1967/68 to 1969/70 relative prices. For a more detailed description, see Duloy and Norton (1973, 1975).

5
Welfare Gains
from Price Stabilization
when Production Is Risky

Price stabilization is one form of intervention by governments in risky markets that has been important and popular over recent decades in many countries and for many commodities. Sometimes schemes labeled as 'stabilization' have been primarily means for transferring assistance to rural producers. Here, however, our concern is with what might be termed 'pure price stabilization.' Our purpose is to explore the welfare implications of complete price stabilization in the particular types of risky markets introduced in Chapter 2. We think of such markets, featuring anticipatory behavior by producers confronted by risky yields (and thus multiplicative risk), as being the typical types encountered in agriculture around the world.

In the literature on the welfare analysis of price stabilization, it is well established that society generally gains from the establishment of costless price stabilization schemes, at least when producers are assumed to be risk neutral.[1] This result was initially demonstrated by Massell (1969) within the context of a linear model of market behavior in which both supply and demand were subject to additive risk terms, and in which producers and consumers were assumed to have perfect information about the market at the time of making their decisions. Subsequent writers extended this finding to nonlinear specifications of the market with multiplicative formulations of the risk terms (Turnovsky 1976, Just et al. 1978) and, initially in the linear case, to the more realistic situation in which producers must act on the basis of price and yield forecasts rather than having perfect information about the market (Turnovsky 1974). More recently, Wright (1979) and Newbery and Stiglitz (1979, 1981) have provided solutions to more general specifications of the market involving lagged or rational expectations, and nonlinear demand and supply functions with multiplicative risks. Newbery and Stiglitz (1981) also specified risk averse producers, and allowed for shifts in supply as producers respond to reduced risks following the introduction of price stabilization. Additional results obtained by Massell (1969), Waugh (1944), and Oi (1961) concerning the distribution of the gain between producers and consumers have proved to be less robust under alternative model specifications, and few generalities have emerged (see Turnovsky (1978), and Newbery and Stiglitz (1981)).

While these findings are of inherent interest, they offer limited guidance as to how large the social gains from price stabilization might be, or about how they might vary with differences in key parameters describing the market structure, including the way in which producers form their price expectations. Some numerical results can be obtained from the algebraic expressions for the social gain offered in the literature, but such results typically apply to over-specialized market structures and do not permit the kind of systematic analysis from which more general quantitative conclusions might emerge. For example, while Newbery and Stiglitz (1981) conclude from their exemplary analysis that the gains from price stabilization largely arise from reduced errors in resource allocation resulting from incorrect price forecasting, they are still compelled to note (p. 23) that 'We have not been able to quantify this effect, except in very simple models.'

In this chapter, we first exploit our partial equilibrium model of agricultural markets and analyze some of the consequences of a complete stabilization of prices. We begin with the nonlinear, multiplicative risk model of a single market developed in Chapter 2, and explore theoretically some qualitative relationships between the way producers form their expectations about prices and yields and the size of the social gain from price stabilization.

In our theoretical analysis we find that society indeed always gains (in terms of expected surplus) from costless stabilization and that, for given $E[P^*]$, the gain is greater the larger is the variability over time in producers' price forecasts. If producers should happen to anticipate the same price each year, the social gain from price stabilization is greater the larger is the deviation of that constant price forecast from expected unit revenue. Indeed, as should be expected from the results in Chapter 2, we find that the social gain from price stabilization will be smallest in a market in which producers use the mean revenue over mean yield as their price forecast in each period.

The significance of the gains and losses from price stabilization cannot be quantitatively assessed very readily in the types of models we use in the theoretical part of this chapter. We therefore resort to a small simulation of the dynamic operation of hypothetical markets. In this way, we show how the size of the social gain from price stabilization, as well as its distribution between consumers and producers, are related to the coefficient of variation in yields, to the elasticities of supply and demand, and to different types of price expectations behavior. We find that the social gain from price stabilization can be quite large when the model is parameterized with empirically plausible values of the coefficient of variation of yield and of the supply and demand elasticities. However, we substantiate Newbery and Stiglitz's (1981) finding that most of gain can be attributed to removing forecasting errors so that, seemingly, the gains from price stabilization can be achieved in large measure by improving producers' forecasting behavior. This result seems of strong policy significance because it is likely that forecasting can be accomplished at lower cost than can buffer-stock stabilization schemes in the event that producers do not hold revenue expectations when making their input decisions.

A limitation of the preceding analyses is that the underlying market specification ignores producers' behavioral response to the reduced risks associated with price stabilization. It also ignores the spillover effects on other markets when the price of one commodity is stabilized. Both aspects were found to be important by Newbery and Stiglitz (1981). Further, and in common with nearly all the literature on price stabilization, our initial analyses ignore policy goals or social consequences other than those captured by changes in producers' and consumers' surpluses. In reality, governments may have very little interest in these surpluses (Cochrane 1980) but they are concerned about the impact of price stabilization on agricultural production, employment, and trade, on the level and distribution of farm income, and on prices and wages. These kinds of considerations are not easily captured in tractable models, so we devote the later part of this chapter to showing how the kinds of agricultural sector models developed in Chapter 4 can usefully be harnessed to addressing this problem. The approach is illustrated by using an agricultural sector model of Guatemala to evaluate a hypothetical bean price stabilization scheme. In the spirit of earlier parts of the chapter, we also examine results for contrasting assumptions about the way in which farmers form their price expectations. Our results reinforce a finding from the simulation model experiments that, as long as producers hold constant price or revenue expectations from year to year, then, for small to moderate values of the coefficient of variation of yield, the gains from price stabilization are not only small but they are about the same for both price and revenue expectations.

PRICE STABILIZATION: SOME THEORETICAL
PROPOSITIONS

To keep our analysis simple, we assume that the only
fundamental source of instability is in yield faced by
producers on the supply side.[2] We thus use the market
structure developed in Equations (2.1) through (2.5) of
Chapter 2. As in Chapter 2, we merely assume convergence. We are specifically interested in two types of
producers' expectations behavior, both of which are
self-fulfilling on average. The first are price expectations, such as the weighted cobwebs

$$P_t^* = \sum_{i=1}^{k} P_{t-i} \gamma_i,$$

which include the Nerlove-type adjustment models. The
second are revenue expectations of the form

$$P_t^* = \sum_{i=1}^{k} P_{t-i} \varepsilon_{t-i} \gamma_i / \mu$$

wherein the covariance between price and yield is captured implicitly.

Measures of Welfare Gain

The measures of welfare we use here are those based
on the surplus concepts introduced in Equations (2.6)
through (2.8). However, for mathematical simplicity, it
will generally be more convenient to work with an alternative but equivalent formulation of (2.8) in which the
inverses of the demand and anticipated supply functions
are used. Let $F = f^{-1}$ and $G = g^{-1}$, then $P_t = F(D_t)$ is
the market demand function and $P_t^* = G(S_t^*/\mu)$ is the
anticipated supply function. Since it is assumed that
the derivatives $f' \leq 0$ and $g' \geq 0$ then $F' \leq 0$ and
$G' \geq 0$. The producers' and consumers' surplus are then

(5.1) $$W_{pt} = P_t S_t - \int_0^{S_t^*} G(Q/\mu) dQ,$$

and

(5.2) $$W_{ct} = \int_0^{S_t} F(Q) dQ - P_t S_t,$$

respectively, and expected social welfare is

(5.3) $$E[W_t] = E[\int_0^{S_t} F(Q)dQ] - E[\int_0^{S_t^*} G(Q/\mu)dQ].$$

Consider the establishment of a buffer stock scheme in which the market price is stabilized at a price $\bar{P}$ which ensures that the buffer stock is self-liquidating on average. Thus, $\bar{P}$ must satisfy $\mu g(\bar{P}) = f(\bar{P})$, and is, therefore, the price corresponding to the intersection of the demand function and the anticipated supply function, that is, $\bar{P} = F(\mu g(\bar{P}))$.[3] Market supply each period is stochastic, $\bar{S}_t = \varepsilon_t g(\bar{P})$, but it is presumed that the stabilizing agency buys the total amount produced each year at price $\bar{P}$, and sells a constant amount $\bar{S} = \mu g(\bar{P})$ to consumers each period at price $\bar{P}$. It is assumed that the scheme is run costlessly and that producers do not expand anticipated supply in response to changed variability of price. The assumption of a costless scheme is clearly not realistic, but is acceptable here because our purpose is only to evaluate the size of the welfare gains from stabilization. Empirical estimates of these costs would have to be obtained if a decision to establish a buffer stock were to be made. Our assumption of no risk-responsive behavior is more restrictive, and we relax this assumption later in formulating a mathematical programming model of the problem.

Under our scheme and using the expected values of Equations (5.1) and (5.2), the producers' and consumers' surpluses are

$$W_{pt} = \bar{P}\bar{S}_t - \int_0^{\bar{S}} G(Q/\mu)dQ$$

and

$$W_{ct} = \int_0^{\bar{S}} F(Q)dQ - \bar{P}\bar{S}.$$

Subtracting the prestabilized surpluses of (5.1) and (5.2), and taking the expected values over t, the gains (or losses) from stabilization accruing to producers and consumers are, respectively,

(5.4) $$E[\Delta W_{pt}] = E[\bar{P}\bar{S}_t - P_t S_t] + E[\int_{\bar{S}}^{S_t^*} G(Q/\mu)dQ]$$

and

(5.5) $$E[\Delta W_{ct}] = E[\int_{S_t}^{\bar{S}} F(Q)dQ] + E[P_t S_t - \bar{P}\bar{S}_t].$$

The social gain is the sum of (5.4) and (5.5),

(5.6) $$E[\Delta W_t] = E[\int_{S_t}^{\bar{S}} F(Q)dQ] + E[\int_{\bar{S}}^{S_t^*} G(Q/\mu)dQ].$$

Some Propositions

Three propositions about the social gain from price stabilization can now be proved.

Proposition 1: <u>The social gain from price stabilization is always positive</u>.

This is proved by first expanding the welfare expression in (5.6) as

(5.7) $$E[\Delta W_t] = E[\int_{S_t}^{E[S_t]} F(Q)dQ] + E[\int_{E[S_t^*]}^{S_t^*} G(Q/\mu)dQ]$$

$$+ \int_{E[S_t]}^{\bar{S}} F(Q)dQ + \int_{\bar{S}}^{E[S_t^*]} G(Q/\mu)dQ.$$

We now proceed to establish that each of these terms is non-negative. By Jensen's inequality,[4]

$$E[\int_{S_t}^{E[S_t]} F(Q)dQ] \geq 0 \qquad \text{since}$$

$$\frac{\partial^2}{\partial S_t^2} \int_{S_t}^{E[S_t]} F(Q)dQ = -F'(Q) \geq 0$$

and

$$E[\int_{E[S_t^*]}^{S_t^*} G(Q/\mu)dQ] \geq 0 \qquad \text{since}$$

$$\frac{\partial^2}{\partial S_t^{*2}} \int_{E[S_t^*]}^{S_t^*} G(Q/\mu)dQ = S'(Q/\mu) \geq 0.$$

Finally, since $E[S_t] = \mu E[g(P_t^*)] = E[S_t^*]$, the sum of the last two integrals in (5.7) can be written as

$$\int_{E[S]}^{\bar{S}} \{F(Q) - G(Q/\mu)\}dQ.$$

Using the mean value theorem, there is a $\tilde{Q}$ within the interval $E[S]$ to $\bar{S}$ such that

(5.8)
$$\int_{E[S]}^{\bar{S}} \{F(Q) - G(Q/\mu)\}dQ$$

$$= \{F(\tilde{Q}) - G(\tilde{Q}/\mu)\}\{\bar{S} - E[S]\}.$$

If $\{\bar{S} - E[S]\} \leq 0$, then $\tilde{Q}$ is greater than $\bar{S}$ and, by assumption, $G'(Q) \geq 0$ and $F'(Q) \leq 0$, so that $G(\tilde{Q}/\mu) - F(\tilde{Q}) \leq 0$ and (5.8) is positive. On the other hand, if $\{\bar{S} - E[S]\} \geq 0$, then $\tilde{Q}$ is less than $\bar{S}$ so that $G(\tilde{Q}/\mu) - F(\tilde{Q}) \geq 0$ and (5.8) is again positive. This completes the requirements for Proposition 1 to hold.

In our second proposition we focus attention on the role of the stability of the price anticipated by producers, and generalize some of the results obtained by Turnovsky (1974).

Proposition 2: For given $E[S_t^*]$, the social gain from price stabilization is larger the greater is the variability of P_t^*.

We prove this, however, by shifting attention to the variability of S_t^*. Since $S_t^* = \mu g(P_t^*)$ is an increasing function of P_t^*, then increasing the variability of P_t^* will increase the variability of S_t^*. It is sufficient, therefore, to prove that the social gain from price stabilization is larger the greater is the variability of S_t^*.

Following Feder (1977), the random variable S_t^* can be written as $S_t^* = u_t + r(u_t - E[S_t^*])$ where u_t is a random variable satisfying $E[u_t] = E[S_t^*]$, and r is a positive scalar. An increase in r has the effect of increasing the variability of S_t^*, but leaves the mean unchanged. Since $S_t = \varepsilon_t S_t^*/\mu$ then $S_t = u_t v_t + r v_t (u_t - E[S_t^*])$ where $v_t = \varepsilon_t/\mu$.

Turning to (5.7), and since only the first two terms are affected by the variability of S_t, we need to show that

(5.9a)
$$\frac{\partial}{\partial r} E[\int_{S_t}^{E[S_t]} F(Q)dQ] + \frac{\partial}{\partial r} E[\int_{E[S_t^*]}^{S_t^*} G(Q/\mu)dQ] \geq 0.$$

Using Leibniz' rule and our transformations of S_t and S_t^* in terms of u_t, (5.9a) can be evaluated as $- E[F(S_t) (u_t - E[S_t^*])v_t] + E[G(S_t^*/\mu)(u_t - E[S_t^*])] \geq 0$. Since $F(S_t) = P_t$ and $G(S_t^*/\mu) = P_t^*$, (5.9a) is equivalent to $E[(P_t^* - P_t v_t)(u_t - E[S_t^*])] \geq 0$, or

$$(5.9b) \qquad cov[P_t^*, u_t] - cov[u_t, P_t v_t] \geq 0.$$

We now establish for (5.9b) that the first covariance is positive and the second is negative under our assumptions. By definition, u_t is positively related to S_t^*, hence $cov[P_t^*, u_t]$ has the same sign as $cov[P_t^*, S_t^*]$. But since $g'(P^*) > 0$ then $\partial S_t^*/\partial P_t^* \geq 0$ and this implies (see note 1, Chapter 2) that $cov[P_t^*, S_t^*] \geq 0$.

A sufficient condition for (5.9b) to hold is now that $cov[u_t, P_t v_t] \leq 0$. Since u_t is positively related to S_t^*, and $v_t = \varepsilon_t/\mu$, then $cov[u_t, P_t v_t]$ has the same sign as $cov[S_t^*, P_t \varepsilon_t]$, and this will be negative if $\partial P_t \varepsilon_t/\partial S_t^* \leq 0$. But $P_t \varepsilon_t = F(\varepsilon_t g(P_t^*))\varepsilon_t$, and hence

$$\partial(P_t \varepsilon_t)/\partial S_t^* = (\partial F(.)/\partial S_t)(\varepsilon_t^2 \partial g(P_t^*)/\partial S_t^*).$$

Now $\partial g(.)/\partial S_t^* = 1/\mu$ and $\partial F(.)/\partial S_t \leq 0$, so that $cov[u_t, P_t v_t] \leq 0$ as required. Thus the inequality in (5.9b) does indeed hold and our Proposition 2 is thus proved.

Proposition 2 takes $E[S_t^*]$ as given. This assumes that $E[P_t^*]$, the mean price expectation over time, is also constant. Thus, for example, the proposition states that the social gain from price stabilization will be smaller if producers adhere to the constant price expectation $P_t^* = E[P_t]$ than if they forecasted last period's price $P_t^* = P_{t-1}$ in the unstabilized market. This is because both price expectations are the same on average, but $P_t^* = E[P_t]$ has zero variability over time, whilst $P_t = P_{t-1}$ has the same variability as P_t. However, the proposition cannot be used to compare the welfare effects of variability in different P_t^* drawn from distributions with different means over time.

By Proposition 2, the social gain from price stabilization will, given the same $E[S_t^*]$, be smallest when producers hold constant price expectations in the unstabilized market. Clearly, not all constant price expectations can be equally good, and it is relevant to search for the constant price expectation, which, if used each period in the unstabilized market, leads to the smallest social gain from the introduction of price stabilization.

Proposition 3: <u>If, in the prestabilized market, producers expect the same price P* in each and every period, then the social gain from price stabilization will be smallest when P* is equal to the expected unit revenue $P^* = E[P_t \varepsilon_t]/\mu$.</u>

Let $A^* = g(P^*)$, then (5.6) can be written as

$$E[\Delta W_t] = E[\int_{\varepsilon_t A^*}^{\bar{S}} F(Q)dQ] + E[\int_{\bar{S}}^{\mu A^*} G(Q/\mu)dQ].$$

Our proposition states that $P^* = G(A^*) = E[P_t \varepsilon_t]/\mu$ is the anticipated price that minimizes $E[\Delta W_t]$. The first-order condition for a minimum is $\partial E[\Delta W]/\partial A^* = -E[F(\varepsilon_t A^*)\varepsilon_t] + G(A^*)\mu = 0$. But $F(\varepsilon_t A^*) = P_t$ given P^*, and $G(A^*) = P^*$, hence $P^* = E[P_t \varepsilon_t]/\mu$ as required. The second-order condition is $\partial^2 E(\Delta W)/\partial A^{*2} = -E[F'(\varepsilon_t A^*) \varepsilon_t^2] + G'(A^*)\mu \geq 0$, or

(5.10) $\mu G'(A^*) \geq E[\varepsilon_t^2 F'(\varepsilon_t A^*)]$.

Since $G' \geq 0$ and $F' \leq 0$, then (5.10) is always satisfied when yields are (necessarily) positive.

We can derive two useful corollaries from Proposition 3. First, since (5.10) is satisfied for all values of A^* and, in turn, P^*, it follows that $E[\Delta W]$ is convex in P^* and that the social gain from price stabilization will be greater the more a constant price expectation P^* deviates from the unit revenue expectation in a market that is not stabilized.

Second, of all possible constant price forecasts in a prestabilized market, the unit revenue expectation has the property of maximizing the welfare function defined in (5.3). This result has, of course, been proved in Chapter 2, but it follows directly here from Proposition 3.

Taking Propositions 2 and 3 together we conclude that, in the absence of exact knowledge about the expected unit revenue $E[P_t \varepsilon_t]/\mu$, then price forecasts should be used which are both unbiased and minimum variance estimators of the expected unit revenue. The lagged revenue forecast $P_t^* = \Sigma_i \gamma_i (P_{t-i} \varepsilon_{t-i}/\mu)$ is unbiased, whereas the lagged price forecast $P_t^* = \Sigma_i \gamma_i P_{t-i}$ is biased. However, the lagged revenue forecast possibly has a larger variance over time and is therefore not necessarily a better estimator than the lagged price forecast.

This completes our theoretical work on the effects of complete price stabilization in our particular, albeit representative, type of risky market. What now seems required for a more comprehensive understanding of the economic issues in such stabilization is quantitative information, first on the magnitudes of the gains and losses--the directions of which were the subjects of our propositions--and, second, on the social costs associated with implementation of stabilization schemes. We now broach the first of these only and shirk presently the second important topic. Fortunately, some work is appearing on the extent of costs (for example, Campbell, Gardner, and Haszler 1980) and on efficient stock-holding rules (see Blandford and Lee 1979, Just and Schmitz 1979), so it may eventually be possible to reach some general conclusions about the economic desirability of stabilization.

GAINS FROM STABILIZATION: SOME QUANTITATIVE INDICATIONS

To explore a quantitative appreciation of our theoretical results, we designed a simulation model of our assumed market structure. Our objectives were to provide a simple but plausible representation of the functioning of a risky agricultural market and to discover orders of magnitude for the gains identified in the theoretical analyses, and their robustness to changes in key parameter values. For modeling simplicity, and in keeping with the theoretical work, we assume complete price stabilization at $\bar{P}$. With this extreme assumption the gains in prospect for the more-typically attempted stabilization schemes involving a band of partially stabilized prices should be exaggerated and perhaps readily identifiable. Needless to say, the size of buffer stock initially on hand to ensure that the market could always clear at $\bar{P}$ would be very large in some cases.

The Simulation Model

The structure of the model can be summarized in four steps. First, functional forms and values of parameters are specified for demand and anticipated (expected) supply and the completely stabilized price $\bar{P}$ is computed. Second, one of seven models of anticipatory or forecasting behavior is specified. Third, a yield is drawn pseudorandomly from an approximate normal distribution of specified mean μ and coefficient of variation $C[\epsilon]$. Fourth, the quantity supplied and market clearing price are computed. Fifth, values of realized prices, revenues, consumers' and producers' surpluses, and the means and variances (over all previous records) of these statistics are computed. Given the specification of steps 1 and 2, steps 3 to 5 are repeated and statistics updated until markets converge to an equilibrium in mean price. As a numerical test of convergence, the program terminates when the mean price for an unstabilized market changes proportionally by less than $C/1000$ between successive periods. In nearly all cases, we found that the mean values of the producers' and consumers' surpluses converged before the mean price; hence, the values of these statistics provided the average gains (or losses) obtained with stabilization as measured under market equilibrium conditions.

The demand and supply functions selected for the reported simulation are both of the constant elasticity form. This simplifying assumption implies that expected price E[P] exceeds the intersection or stabilized price $\bar{P}$ and has the parametric economy of being specified only in terms of the constant elasticities. While the choice of any particular functional form obviously limits the generality of the results, the representation accords with that frequently used by empirical analysts and, in this sense, might be indicative of more general empirical specifications.

For any run, two groups of initial conditions were required. Where values of lagged endogenous variables were required in some expectations models, they were taken to be initially as the computed $\bar{P}$. A seed for the pseudorandom generator was drawn for each run. The second group consisted of the market parameters: elasticity of demand (expressed as a positive number, e_d; elasticity of supply, e_s; and coefficient of variation of yield, $C[\varepsilon]$. An experimental design was used to explore the influence of these parameters over a range of more-or-less plausible values. A complete factorial design including all 27 combinations of three levels of the three factors was used, namely, e_d = 0.5, 1, 1.5; e_s = 0.5, 1, 2; $C[\varepsilon]$ = 0.1, 0.25, 0.4. Gains from stabilization are measured by changes in consumers' surplus and producers' ex post surplus as defined in Equations (5.4) and (5.5). The surplus measures are computed by integration of areas under the respective curves although, in the case of the assumed constant-elasticity demand curve, it is necessary for measuring consumers' surplus to impose a lower quantity bound on the range of integration. This bound was located at a quantity corresponding to four times the stabilized price $\bar{P}$, which is a fixed value for any given combination of e_d and e_s.

Results from the Simulation Experiments

Some results are presented for the design points e_d = 0.5, e_s = 0.5, $C[\varepsilon]$ = 0.1, 0.25, and 0.4 in Table 5.1. These include estimates of the gains from stabilization for seven alternative supply price specifications: (i) producers anticipate the intersection price ($P^* = \bar{P}$), (ii) expected price ($P^* = E[P]$), (iii) past year price ($P_t^* = P_{t-1}$), (iv) a weighted average of past prices ($P_t^* = 0.5\,P_{t-1} + 0.3\,P_{t-2} + 0.2\,P_{t-3}$), (v) past year revenue ($P_t^* = P_{t-1}\varepsilon_{t-1}/\mu$), (vi) a weighted average of past revenues (with weights as in (iv)), and (vii) expected revenue ($P^* = E[P\varepsilon]/\mu$).

TABLE 5.1
Welfare Gains from Stabilization for Alternative Expectation Models and Levels of Yield Variability (Supply and demand elasticities both 0.5)

Expectations Model	Coefficient of Variation of Yield	Gains from Stabilization as Percentage of Consumers' Expenditure in the Stabilized Markets			Gains Due to Removing Forecasting Error as Percentage of Consumers' Expenditure in Stabilized Markets		
		Producers	Consumers	Total	Producers	Consumers	Total
(i) Intersection price forecast, $P^* = \bar{P}$	0.1	0.778	0.974	1.752	2.99	-2.91	0.07
	0.25	-27.142	36.382	9.237	-9.25	9.56	0.31
	0.4	-5.089	27.800	22.711	-2.31	2.17	-0.14[a]
(ii) Mean price forecast, $P^* = E[P]$	0.1	-0.762	2.236	1.474	1.45	-1.65	-0.21[a]
	0.25	0.670	11.855	12.526	18.56	-14.96	3.60
	0.4	25.716	3.450	29.166	28.50	-22.18	6.32
(iii) Naive lagged price forecast, $P^*_t = P_{t-1}$	0.1	15.495	14.578	30.073	17.70	10.69	28.39
	0.25	49.347	34.045	83.392	67.24	7.23	74.47
	0.4	71.294	39.690	110.986	74.08	14.06	88.14
(iv) Weighted lagged price forecast, $P^*_t = f(P_{t-i})$	0.1	0.038	2.251	2.293	2.25	-1.64	0.61
	0.25	-0.23	16.194	15.963	17.66	-10.63	7.04
	0.4	23.986	13.133	37.119	26.77	-12.50	14.27
(v) Naive lagged 'revenue' forecast, $P^*_t = R_{t-1}$	0.1	2.672	7.368	10.040	4.88	3.48	8.36
	0.25	0.996	31.370	32.366	18.89	4.55	23.41
	0.4	9.838	36.452	46.292	12.62	10.82	23.45
(vi) Weighted lagged 'revenue' forecast, $P^*_t = f(R_{t-i})$	0.1	-2.470	3.874	1.404	-0.26	-0.01	-0.28[a]
	0.25	-21.280	32.597	11.318	-3.39	5.78	2.39
	0.4	-13.541	39.250	25.706	-10.76	13.62	2.86
(vii) Expected 'revenue' forecast or rational price expectation, $P_t = E[R]$	0.1	-2.207	3.886	1.679	0.00	0.00	0.00
	0.25	-17.892	26.819	8.927	0.00	0.00	0.00
	0.4	-2.783	25.633	22.847	0.00	0.00	0.00

[a] These figures are slightly negative due to convergence difficulties in the measures of expected social gains.

We now overview the tabular results by reference to some statistics from the experiment. For the sake of brevity, only a 1/9 portion of the results is reported in Table 5.1. The first result apparent in the simulations is that the average welfare gains or losses are generally considerable and tend to be quite large for the case of high relative variation (coefficient of variation equal to 0.4).

There are two ways to look at the gains (or losses) of consumers and producers from stabilization. First, the surpluses can be viewed as compensating variations-- in the case of the positive gains, as the amount of money that the consumers, the producers or both would be willing to pay to enact a complete stabilization scheme. Second, some form of relative benefits can be considered such as, for example, the gains as ratios of the associated stabilization costs.

Because stocks and costs were not modeled, we adopted the procedure of expressing gains as proportions of total expenditure in the completely stabilized market. Such a procedure generates both an absolute and relative measure of gains in the following sense. Given a percentage gain, the absolute level of the compensating variation for each group of agents can be computed for a market of any size having the same elasticity and stochastic characteristics. This can be done by simply multiplying the percentage gains as given in Table 5.1 by an estimate of the total expenditure in the market of interest. Stabilized total expenditure represents the value of the market transactions that would occur under a complete stabilization and is used in this case as a measure of market size. For this purpose, it has the added advantage of remaining constant for given values of the market parameters, regardless of the type of price expectations held by producers in the unstabilized market.

From an alternative point of view, the percentage gains are obviously a relative measure of benefits from stabilization. For each 100 dollars transacted in the stabilized market, they represent how much the consumers, the producers or both would be willing to pay, on average, to enact the stabilization scheme.

In relative terms, these gains range from a few percentage points, for the cases of low elasticities and/or low relative variation, to more than 100 percent, for the more extreme cases of naive lagged expectation models. In absolute terms, these percentage gains imply, for example, that in a market of the size of the international wheat market (about 16 billion dollars of recorded transactions), depending on the type of expectations held by producers, costless stabilization could achieve gross gains ranging from a few hundred million dollars to several billion dollars. Perhaps this is why

there has been such enthusiasm in some quarters for
_global stabilization.

Such a conclusion, though seemingly favorable for
stabilization, is tempered by the observation that, in
many of our simulation runs, a large proportion of the
social gain from stabilization occurs because producers
make improper price forecasts. Proposition 2 predicts
that the social gain will be larger the greater the
variability in P_t^*, but the numerical importance of the
proposition can be quite surprising. For example, even
given a low value of the coefficient of variation of
yields of 0.1, the social gain from price stabilization
is 30 percent of consumers' expenditure in the stabilized market if producers plan on the basis of last
period's price (that is, $P_t^* = P_{t-1}$). But the gain is
only 1.5 percent of stabilized expenditure if producers
plan on the basis of the mean equilibrium price ($P_t^* = E[P]$). Since $E[P^*]$ is the same in both cases, the
large difference in the relative welfare gains is due
solely to differences in the variability of P_t^*. A similar result holds with revenue expectations. Again
selecting $C = 0.1$, the social gain from stabilization is
10 percent of stabilized expenditure when producers
expect last period's unit revenue ($P_t^* = R_{t-1}/\mu$), but is
only 1.7 percent when $P_t^* = E[R]/\mu$. Again $E[P^*]$ is the
same in both cases, and the larger social gain from stabilization arises when the variability in P_t^* is non-
zero. Of course, one-year lagged price or revenue
expectations are exceedingly naive and have high variability over time. The weighted lagged forecasts
(expectations models (iv) and (vi) in Table 5.1) are
considerably less variable over time and, for example,
when $C = 0.1$ and $P_t^* = f(P_{t-i})$, the social gains are not
much larger than the gains obtained with constant price
expectations. However, such relative differences
translate into significant money amounts in realistically sized markets.

By Proposition 3, the social gains from price stabilization must be smallest when producers plan on the
basis of expected unit revenues. Consequently, the
social gains reported for expectations model (vii) in
Table 5.1 are the smallest gains possible for each value
of C over all other expectations models. Larger gains
than these are directly attributable to inferior price
forecasting behavior in the unstabilized market, and
those additional gains could be obtained by improving
producers' forecasting behavior without setting up a
price stabilization agency. The right-most three col-
umns of Table 5.1 show the differences between the gains
with each of the different expectations models and the
gains with the expected unit revenue expectation. These
figures measure directly the gains and losses arising
from the removal of inferior forecasting behavior.

They are, of course, zero for the expected unit revenue expectation but, in many cases, the largest part of the stabilization gain can be achieved merely by improving producers' price forecasts. This is particularly so in the cases where producers plan on the basis of lagged prices or revenues. Surprisingly though, $P_t^* = \bar{P}$ -- the price corresponding to the intersection of demand and anticipated supply -- performs about as well as the expected unit revenue expectation. The expected price forecast $E[P]$ also performs well for small to moderate values of the coefficient of variation C.

With such improved forecasting, however, the distribution of benefits between producers and consumers could change substantially. For the two cases of expected price and weighted lagged price expectations, for example, a strategy of improved forecasts would shift from consumers to producers a considerable part of the gains that might be achieved via a buffer stock policy. In most other cases, the transfers between consumers and producers via stabilization are either zero or small. With the exception of the intersection price model and the naive model, consumers and producers either both gain from stabilization or both lose insignificant amounts.

In our discussion of results so far we have concentrated on those for the inelastic markets reported in Table 5.1. Market structures of great diversity were included in the complete experiment and we turn now to explore what generalizations are possible when a wide range of elasticities is considered. To overcome the difficulty and tedium of reporting a great bulk of results we elected to summarize the information by means of some regression equations. We took the gains to producers, consumers and society expressed as percentages as in Table 5.1 as separate dependent variables and, for each model of expectations behavior, related these in least-squares regression to a complete second-order response model (that is, with intercept, linear, quadratic and interaction terms) in the three experimental factors, namely e_d, e_s and $C[\varepsilon]$.

The regressions are not reported because of their bulk.[5] They permit prediction of relative gains at any specified point within the design space and description (via partial differentiation) of the relative changes in welfare with respect to changes in the parameters of the market. The partial derivatives of a complete second-order surface are functions of all the factors so these are evaluated at an arbitrary point for the purpose of our discussion. To complement the statistics in Table 5.1, we present in Table 5.2 the marginal effects as evaluated at the arbitrary point: $e_d = 0.5$, $e_s = 0.5$, and $C = 0.2$.

TABLE 5.2
Marginal Changes in Relative Welfare with Respect to Key Parameters
of the Risky Market

Expectations Model (See Table 5.1)	Change in Percentage Gain to Respective Groups for a Small Change in Respective Parameters[a]								
	Consumers w.r.t.			Producers w.r.t.			Total w.r.t.		
	e_d	e_s	C	e_d	e_s	C	e_d	e_s	C
(i) $P^*_t = \bar{P}$	-28	-3	79	17	1	-40	-11	-2	41
(ii) $P^*_t = E[P]$	-14	-9	17	-14	7	33	-15	-2	50
(iii) $P^*_t = P_{t-1}$	-60	1	49	-191	366	403	-251	454	676
(iv) $P^*_t = f(P_{t-i})$	-9	-7	24	-72	26	62	-81	21	84
(v) $P^*_t = R_{t-1}$	-57	10	75	31	46	-278	-26	55	-203
(vi) $P^*_t = f(R_{t-i})$	-17	-11	72	-83	15	6	-99	13	76
(vii) $P^*_t = E[R]$	-20	-2	60	9	0	-21	-12	2	40

a Evaluated at $e_d = 0.5$, $e_s = 0.5$, $C = 0.2$ after partially differentiating the response functions.

We found in algebraic manipulation of our underlying
model and of expressions for the changes in surpluses
that it was not possible to sign these marginal effects.
Our failure to do so is supported by the numerical
results which show sign changes of marginal effect
across expectations models and, naturally, across
affected groups in society.

In spite of the variability of these results, a few
generalizations can be braved, especially if the relatively poor expectations models (like the naive cobweb
models) are underemphasized. Consumers' gains are most
sensitive to changes in C and least sensitive to changes
in e_s. Consumers' gains from stabilization increase
with increasing C, decrease with increasing e_d (that is,
as demand becomes less inelastic) and tend to decrease
with increasing e_s. Producers' gains from stabilization
are affected by changes in e_d, e_s and C in a generally
ambiguous manner depending on the nature of their expectations. The gains to society as a whole tend to change
systematically with respect to changes in C and e_d,
being most sensitive to the former. They typically increase with increasing C (as we would expect from our
Proposition 2) and decrease with increasing e_d (less
inelastic demand).

Conclusions from the Simulation Experiments

What are the practical implications of these results? First, because of the range of coefficients of
variation used, the estimated gains from stabilization
should give a reasonable idea of the gains that might be
obtained by stabilizing prices in typical agricultural
markets. Our results show that these gains can be quite
large. For example, if demand and supply elasticities
are both 0.5 and the coefficient of variation of yields
is 0.1, then for each $200 transacted in the stabilized
market (or each ton of wheat, say) the social gains from
price stabilization range from $3 given expected unit
revenue expectations to $60 when producers expect last
period's price. These gains increase to $18 and $167
respectively, when the coefficient of variation is 0.25.
If producers plan on the basis of a weighted average of
past prices or revenues, as is commonly assumed in
empirical supply analysis, then the social gain from
price stabilization will take on intermediate but still
sizeable values.

Our analysis has been based on the assumptions of
costless storage and complete price stabilization, hence
these estimates of the social gains must be interpreted
as the maximum gross gains attainable from price stabilization. Nevertheless, there are clearly some plausible market conditions under which more realistically

designed stabilization schemes could return a substantial net social benefit.

Second, a program of data collection, appropriate forecasting, and information dissemination could achieve a large part of the gains of a buffer stock scheme in situations where producers act on the basis of price forecasts different from the expected per unit revenue expectation. Such a market information service should provide producers with an estimate of the expected unit revenue, given the structural parameters of the market. Our simulation results also show that the intersection price of demand and anticipated supply is about as good a price forecast, as is the expected price for low to moderate values of the coefficient of variation of yield. If producers consistently use any of these price forecasts, unless the coefficient of variation of yields is exceptionally high, price stabilization is unlikely to return an attractive gain in terms of the social welfare measure used in this paper.

Finally, it is also worth noting that, in many of the cases considered, the period of convergence in mean market price was considerable (say about 50 periods). Further gains from stabilization might result from the fact that the short-term variance of prices is only slowly reduced to its long-run value in an unstabilized market.

EVALUATING PRICE STABILIZATION WITH MATHEMATICAL PROGRAMMING

Stabilization interventions in any one market may have important spillover effects in other markets that are not captured in the kinds of single-commodity analyses reported above. These spillover effects may arise from commodity substitution in demand, and/or from competition for scarce resources in production between the stabilized crop and other commodities. In addition, even though the importance of producers' aversion to price and yield risks has been recognized as a determinant of supply (Just 1974), the implied changes in average supply following price stabilization typically are ignored.[6] Also, nearly all the analytical work on price stabilization has focused on a very narrow set of policy objectives, particularly the changes in producers' and consumers' surplus, producers' income, and storage costs. However, price stabilization may affect a wider range of policy issues when allowance is made for risk response and multimarket interactions.

Price-endogenous mathematical programming models of the kind described in Chapter 4 can take account of multiproduct relationships in supply and demand, and can simulate the effects of risk-averse behavior at the farm

level. They also can provide a wealth of detailed
information about production, resource use, consumption,
prices, and trade, at both the micro (farm or regional)
and sectorwide levels. In this section we show how
these models can be used to evaluate price stabilization
schemes. The method is illustrated using an agricultural sector model of Guatemala to evaluate a hypothetical bean price stabilization scheme.

An Agricultural Sector Model of Guatemala

The Guatemalan model used here is typical of the
price-endogenous mathematical programming models
described in Chapter 4. In particular, the model has a
linear and integrable demand system, a linear constraint
set, and a risk behavior specification of the mean,
standard deviation type.

The model is fully described in Pomareda (1980), and
only a brief summary is attempted here. Production is
modeled by three different farm sizes representing primarily subsistence farms, market-oriented farms which
rely predominantly on family labor, and large commercial
farms which use hired labor and produce most of the traditional export crops. Each farm type has a linear
programming submatrix in which monthly constraints on
land and labor are represented, as well as minimum
constraints on food crop production. The hiring and
selling of family labor between farm groups is per-
mitted. As in Chapter 4, farmers are assumed to maximize a utility function defined in terms of the mean and
standard deviation of income. Specifically, this function takes the form $u = E[y] - \phi \sigma_y$ where y denotes
income, and ϕ is a risk aversion parameter.

The production of different crops (maize, rice, sorghum, beans, wheat, cotton, bananas, sugar and coffee)
is aggregated to the national level and transformed
through processing activities in the model to final products (for example, wheat flour) for the domestic and
export markets. Domestic demand is represented by the
linear demand system $p = a - Bq$, where p denotes the
$n \times 1$ vector of domestic retail prices, a and B are
$n \times 1$ and $n \times n$ matrices of demand coefficients, respectively, and q is the $n \times 1$ vector of quantities
demanded. The quantities demanded are defined to
include all food retained by farm families as well as
usual market transactions, a specification which permits
farm family food consumption to be valued at market
prices. The matrices of demand parameters, a and B, are
calculated in a way which reflects this measure of total
demand. Trade barriers (sugar quotas, for example) are
incorporated in the model, and tariffs and other trade

costs are charged to the trade activities where appropriate.

To simplify notation, our algebra abstracts from the farm type differentiations incorporated in the model, and assumes that all commodities are traded internationally. In fact, trade activities are included in the model only to permit food crop imports and nonfood crop exports. These restrictions closely approximate past Guatemalan trade policies, and can lead to important differences between domestic and world prices.

Assuming price expectations behavior, the model objective function that provides the competitive solution to prices and quantities in all markets is:

(5.11)
$$\text{Max}_{x} \, U = E[q'] \, (a - 0.5 \, B \, E[q]) - c_x'x$$
$$+ \, c_e'e - c_m'm - \Phi(x'\Gamma x)^{0.5}$$

where $E[q] = E[N] \, x + m - e$, and x is an n x 1 vector of crop areas grown; N is an n x n diagonal matrix of stochastic yields; m and e are n x 1 vectors of tons of imports and exports, respectively; c_x is an n x 1 vector of production costs per unit area; c_m is an n x 1 vector of import costs per ton; c_e is an n x 1 vector of export prices per ton net of export costs; Γ is an n x n covariance matrix of crop revenues (prices times yield);[7] and Φ is a suitable average of individual farmers' risk-aversion parameters. This maximand is an elaboration of Equation (4.21), and is only applicable if B is symmetric (the integrability requirement). In the absence of estimates of cross-price effects, the B matrix is specified as diagonal in the Guatemalan model.

The alternative model objective function which assumes revenue expectations is, following Equation (4.22),

(5.12)
$$\text{Max} \, E[W] = E[q'(a - 0.5 \, Bq)] - c_x'x$$
$$+ \, c_e'e - c_m'm - \Phi(x'\Gamma x)^{0.5}$$
$$= \text{Equation (5.11)}$$
$$- \, 0.5 \, \sum_{ij} \sum x_i \, x_j \, \sigma_{ij} \, b_{ij}$$

where σ_{ij} denotes the covariance between the yields of the ith and jth crops.

In Chapter 4, we treated the covariance matrix of crop revenues as a constant, and assumed that it is typically estimated on the basis of time-series data on prices and yields. While observed price and yield deviations around their mean (or trend lines) may be an acceptable measure of risk in market equilibrium, the relevant elements of Γ are not invariant with respect to their mean prices.[8] Thus, if the expected prices in the equilibrium solution are different from the sample mean prices used in the calculation of Γ, then Γ should be revised. Procedures for endogenizing the Γ matrix have not yet been developed; therefore, in the Guatemalan model, we used an iterative procedure. If, at the tth iteration the i,jth element of Γ had mean prices $\bar{p}_{it}$ and $\bar{p}_{jt}$, and these differed from the equilibrium prices $E[p_{it}]$ and $E[p_{jt}]$ obtained in the corresponding tth model solution, then $E[p_{it}] - \bar{p}_{it}$ and $E[p_{jt}] - \bar{p}_{jt}$ were added to the sample price observations for the ith and jth crops, the relevant i, jth element of Γ was recalculated, and a new solution was obtained. This procedure was repeated until $E[p_{it}] - \bar{p}_{it}$ and $E[p_{jt}] - \bar{p}_{jt}$ converged to approximately zero for all the elements of Γ. In practice, Γ typically converged in three or four iterations.

Methodology of Price Stabilization Experiments

We are interested in a stabilization scheme in which the domestic price of beans is fixed at its expected market equilibrium value. Such price stabilization would be achieved through the establishment of buffer stocks. To assure a self-liquidating stock on average, the price at which a market is to be stabilized is the expected market clearing price in equilibrium. This price can be obtained from the model. The problem is to modify the model to obtain the market equilibrium solution corresponding to the stabilized situation.

The model solutions are conditioned in part by the covariance matrix Γ, and stabilizing the price of the jth crop changes the variance and the covariance terms involving that crop. An important part of the method of experimenting with price stabilization therefore follows: one must recalculate all the relevant elements of Γ using the stabilized price $\bar{p}_j = E[p_j]$ and then resolve the model for a new equilibrium.

However, producers will adjust their cropping patterns to arrive at a new optimal plan given their assumed $E[y]$, σ_y utility functions. This is the risk-response effect induced by stabilization, and the original expected market-clearing price for the stabilized crop no longer will be the same. The stabilized price, $\bar{p}_j = E[p_j]$, now will have to be revised to retain a self-liquidating buffer stock, the element of Γ recalculated, and the solution process repeated. This iterative procedure is repeated until Γ converges.

As will later become clear, a small modification is also required in the demand specification for the stabilized crop in Equation (5.12). This is because q_j is no longer stochastic when the stabilizing agency sells a fixed amount $\bar{q}_j$ to consumers each year, and Equation (5.12) must be revised so that $E[q_j] = \bar{q}_j$ for the stabilized commodity. This can be done by setting $\sigma_{ij} = 0$, all i, in Equation (5.12).

The poststabilized solution provides the expected values of all activities in the new market equilibrium. Any changes from the prestabilized solution stem from supply adjustments following changes in farm-level risks, or from the disappearance of the covariance between price and yield leading to identical revenue and price expectations for the stabilized crop. Assuming producers were risk-neutral ($\Phi = 0$) and that they plan on the basis of price expectations (objective function (5.11)), then the pre- and poststabilized solutions would, in fact, be identical, with $\bar{p}_j$ remaining constant. Even though the model activity levels would not change under these conditions, the removal of price and market supply variations still leads to changes in the expected values of the consumers' surplus, producers' surplus, and income.

The surplus and income changes can be calculated in the model. Given our assumed market structure, prices in the tth year are given by $p_t = a - Bq_t$.

Expected consumers' surplus in the prestabilized situation is

(5.13)
$$E[W_{ct}] = E[q_t'(a - 0.5\ Bq_t) - p'q_t]$$

$$= 0.5\ E[q_t'Bq_t],$$

and the ex post expected producers' surplus (see also Equation (5.4)) is

(5.14)
$$E[W_{pt}] = E[q_t'(a - Bq_t)] - c_x'x + c_e'e$$

$$- c_m'm - \Phi(x'\Gamma x)^{0.5}.$$

The producer's surplus is defined net of the risk term $\Phi(x'\Gamma x)^{0.5}$, which is the income compensation that producers require for accepting the risks associated with x. By deleting this term in Equation (5.14), the expected value of producers' income in the prestabilized markets is obtained.

To measure aggregate social welfare, we again use the sum of the expected producers' and consumers' surpluses. In the prestabilized market, this is the sum of Equations (5.13) and (5.14) and is equal to model objective function (5.12).

The establishment of a buffer stock agency stabilizes the prices of a subset of the vector p_t. By partitioning the relevant matrices, the price and quantity vectors are

(5.15)
$$\begin{bmatrix} p_{1t} \\ p_{2t} \end{bmatrix} = \begin{bmatrix} a_1 \\ a_2 \end{bmatrix} \begin{bmatrix} B_{11} & B_{12} \\ B_{21} & B_{22} \end{bmatrix} \begin{bmatrix} q_{1t} \\ q_{2t} \end{bmatrix} , \text{ and}$$

(5.16)
$$\begin{bmatrix} q_{1t} \\ q_{2t} \end{bmatrix} = \begin{bmatrix} N_{1t} & 0 \\ 0 & N_{2t} \end{bmatrix} \begin{bmatrix} x_1 \\ x_2 \end{bmatrix} + \begin{bmatrix} m_1 \\ m_2 \end{bmatrix} - \begin{bmatrix} e_1 \\ e_2 \end{bmatrix} .$$

If the buffer stock agency wishes to stabilize prices p_1 at $\bar{p}_1$, where $\bar{p}_1$ is the vector of prices that ensure self-liquidating stocks on average, the agency would plan to buy all the production of q_1 each year and, by controlling imports m_1 and exports e_1, to release the quantities of q_1 to the domestic market each year that are required to maintain prices at $\bar{p}_1$. If q_1 and q_2 are demand-independent groups (that is, $B_{21} = B_{12} = 0$), the agency trades constant amounts $\bar{m}_1$ and $\bar{e}_1$ each year and sells constant amounts $\bar{q}_1 = E[N_1]X_1 + \bar{m}_1 - \bar{e}_1$ to the domestic market.[9] When q_1 and q_2 are not demand independent, p_1 is subject to random variations arising from q_2 as well as from variations in q_1. Since the agency would not carry stocks of q_2, the actual quantities of q_1 sold would have to be varied from year to year to compensate for variations in q_2. $\bar{q}_1$ would then denote only the expected value of the amount sold by the agency to maintain prices at $\bar{p}_1$.

Where $\bar{q}_1$, $\bar{m}_1$ and $\bar{e}_1$ are not stochastic variables, the expected consumers' and producers' surpluses in the stabilized situation are

(5.17)
$$E[W_c] = 0.5 \ \bar{q}_1' B_{11} \bar{q}_1 + 0.5 \ E[q_2' B_{22} q_2],$$

and

(5.18)
$$E[W_p] = \bar{q}_1'(a_1 - B_{11}\bar{q}_1) + E[q_2'(a_2 - B_{22}q_2)]$$
$$- c_x'x - c_m'm + c_e'e - \Phi(x'Tx)^{0.5}.$$

Expected producers' income is then Equation (5.18) with the risk term $\psi(x'\Gamma x)$ omitted.

Taking the sum of Equations (5.17) and (5.18), expected social welfare in the stabilized situation is

$$(5.19) \qquad E[W] = \bar{q}_1'(a_1 - 0.5\, B_{11}\bar{q}_1)$$

$$+ E[q_2'(a_2 - 0.5\, B_{22}q_2)]$$

$$- c_x'x - c_m'm + c_e'e - \psi(x'\Gamma x)^{0.5}.$$

Equation (5.19) is a modified version of (5.12) in which $E[q_j^2]$ is replaced by $\bar{q}_j^2$ for all stabilized commodities. It is also the relevant model maximand for obtaining the market equilibria corresponding to revenue-forecasting behavior in the stabilized situation. Because prices and yields are no longer correlated for the stabilized commodities, producers act as price forecasters when planning x_1 and as revenue forecasters when planning x_2.

The gain in expected social welfare from stabilizing p_1 is the value of Equation (5.19) minus Equation (5.12). If producers are risk-neutral and plan on the basis of price forecasts, the values of x, m, and e remain constant; $\bar{q}_1$ equals $E[q_1]$ of the prestabilized situation; and the welfare gain is

$$(5.20) \qquad E[\Lambda W] = 0.5(E[q_1'B_{11}q_1] - \bar{q}_1'B_{11}\bar{q}_1)$$

$$= 0.5 \sum_{ij} x_{1i}\, x_{1j}\, \sigma_{1ij}\, b_{1ij}.$$

To obtain the values of the surpluses and income measures defined above, it is only necessary to incorporate Equations (5.11), (5.12), and (5.14) into the model and to have access to the value of $\psi(x'\Gamma x)^{0.5}$ from the solution. Since either (5.11) or (5.12) would be the model maximand, then only two additional accounting rows are required. These are quadratic equations, but they can be linearized concurrently with the objective functions using the techniques described in Chapter 4.[10]

An assumption underlying the derivation of these welfare measures is that the activity levels x, m, and e are nonstochastic. Since x (the crop areas planted) depends in part on producers' forecasts about prices, it is implied that producers hold constant forecasts over time.

If the assumption of nonstochastic activities is relaxed, the surplus and income measures used in the model will be incorrect. It is possible to derive the correct measures (see Hazell and Pomareda 1981), but they involve additional terms that cannot be calculated in a mathematical programming model. Our present approach to evaluating price stabilization schemes is therefore limited to situations where we are willing to assume that producers anticipate either expected price or expected revenue every period. Evaluating the gains from stabilization under this assumption is still useful though since, as we saw from the simulation experiments, these are exactly the situations where any realizable gains from price stabilization are additional to the gains that can be realized from simply reducing the variability of producers' price or revenue forecasts. In other words, these are the gains that might potentially warrant the establishment of a price stabilization scheme rather than more simply improving market information services.

The Guatemalan Experiments

Table 5.3 contains some results for selected food crops obtained from the model, together with actual data for 1976, the year for which the model was numerically specified. The results are presented for price and revenue forecasting behavior (using model maximands (5.11) and (5.12), respectively) and for three different levels of risk aversion. A Φ value of zero implies risk neutrality; Φ values of 1.65 and 3.16 represent 'reasonable' and 'extreme' levels of risk aversion, respectively.

The results in table 5.3 suggest that the model describes 1976 production levels quite well. They are consistent with an assumption of 'reasonable' risk behavior ($\Phi = 1.65$). The results obtained for price and revenue expectations are similar for given values of Φ. As risk aversion increases, bean production is significantly curtailed. Beans are clearly a high-risk crop and a suitable candidate for illustrative price stabilization experiments.

Price stabilization cannot affect the model's activity levels if producers are risk neutral and plan on the basis of price expectations, but it does lead to a small gain in social welfare of about $5 million (or $50/ton of beans produced). The more interesting results for Φ values of 1.65 and 3.16 are summarized in Table 5.4.

Price stabilization for beans leads to a gain in social welfare of 1.3 percent (or $12.75 million) when Φ = 1.65 and producers hold constant price expectations. This is equivalent to a gain of $135/ton of beans

TABLE 5.3
Results for Various Levels of Risk Aversion and
Alternative Expectations Behavior, Guatemala

	Price Expectations			Revenue Expectations			1976 Actuals
	$\Phi = 0$	$\Phi = 1.65$	$\Phi = 3.16$	$\Phi = 0$	$\Phi = 1.65$	$\Phi = 3.16$	
Production (10^3 tons)							
Maize	1076.5	1031.6	1031.6	1076.5	1031.6	986.8	1005.7
Rice	27.9	27.9	29.0	26.7	26.7	26.7	34.1
Sorghum	52.3	50.2	48.1	50.2	48.2	46.0	49.2
Beans	98.1	90.2	41.2	93.5	86.3	41.2	92.1
Wheat	67.1	67.1	67.1	67.1	67.1	67.1	56.9
Social Welfare (millions $U.S.)	1067.0	980.9	898.0	1071.9	981.6	898.7	

produced which, while sizeable, is probably too small to justify the costs of a buffer stock. This gain is larger than the results obtained from the simulation model and reported in Table 5.1. The difference arises partly because producers are now assumed to be risk averse, and because the demand for beans in the Guatemalan model is highly inelastic.

As required by Proposition 3, the gain is smaller when producers hold revenue expectations -- it is $12.0 million, or 5.9 percent smaller. The smallness of the difference is consistent with our simulation results in Table 5.1 which showed that the expected price and revenue forecasts were about equally efficient for small to moderate values of the coefficient of variation of yields. The gain from stabilization is larger if producers are assumed to be more risk averse; it increases to $15 million when $\Phi = 3.16$, or about $159 per ton.

There are other effects from stabilizing bean prices. When $\Phi = 1.65$, bean production increases by 4.3 percent and 9 percent for price and revenue expectations, respectively. This additional output is produced with resources that otherwise would be idle.[11] There is also a sizeable decline in the standard deviation of

TABLE 5.4
Results for Various Price Stabilization Experiments, Guatemala

	Φ = 1.65		Φ = 3.16		Φ = 1.65	
	Prestabilized Price Model	% Change with Bean Price Stabilized	Prestabilized Price Model	% Change with Bean Price Stabilized	Prestabilized Revenue Model	% Change with Bean Price Stabilized
Income and Welfare Measures (millions $US)						
Social welfare	980.9	1.30	898.0	1.65	981.6	1.22
Consumers' surplus	906.3	0.02	896.9	-2.80	891.6	0.87
Producers' income	275.3	2.36	263.2	17.10	288.9	-0.12
Standard deviation of producers' income[a]	53.7	-7.60	49.4	1.77	53.1	-6.54
Agricultural Trade Balance (millions $US)	281.1	0	260.3	7.88	282.2	0
Agricultural Employment (Thousands full-time jobs)	5083.0	0.33	4946.8	1.34	5059.5	0.65
Production (10^3 tons)						
Maize	1031.6	0	1031.6	-4.34	1031.6	0
Rice	27.9	0	29.0	-3.79	26.7	0
Sorghum	50.2	0	48.1	0	48.2	0
Beans	90.2	4.32	41.2	118.93	86.3	9.04
Wheat	67.1	0	67.1	0	67.1	0
Retail Prices ($US/ton)						
Maize	179	0	179	14.53	179	0
Rice flour	322	0	258	24.81	387	0
Sorghum	171	0	199	0	199	0
Beans	516	-14.15	590	-12.54	590	-24.92
Wheat flour	510	0	510	0	510	0

a Sum of standard deviations over all farm groups.

producers' income. In both cases, the domestic price declines substantially, and there is an increase in agricultural employment, of 17,000 jobs in the case of price expectations, and of 33,000 jobs with revenue expectations. The gains are more exaggerated under extreme risk aversion, and bean production more than doubles. However, because beans are imported in the prestabilized solution for this value of Φ, the extra production largely substitutes for imports. This leads to a decline in the domestic price of only 13 percent. The agricultural trade balance is affected favorably, and the large increase in bean production leads to a decrease in maize and rice production and an increase in the standard deviation of producers' income.

The results in Table 5.4 show some ambiguity in the gain to producers and consumers. Consumers gain from bean price stabilization when $\Phi = 1.65$, but lose when $\Phi = 3.16$. Average producers' income increases when they hold price expectations (by 2.3 percent and 17 percent for Φ values of 1.65 and 3.16, respectively) but declines when they plan on the basis of revenue expectations.

CONCLUSION

In this chapter we have established that the way in which farmers form their price expectations has a very important bearing on the size of the social benefits that can be realized from price stabilization schemes. If, in the prestabilized market, producers hold naive lagged price or revenue expectations, the social gain from price stabilization may be large. However, if producers initially forecast expected price or revenue, or anticipate the intersection price between demand and expected supply, the social gain is much more modest. This finding reinforces work by Newbery and Stiglitz (1981), and strongly suggests that, where price stabilization is found to be socially beneficial, the larger part of the gain might more easily be obtained through a program of data collection, appropriate forecasting, and information dissemination for producers.

In arriving at these results we have attempted to refine and sharpen the available tools of analysis to give more adequate representation to producers' risk responsive behavior, to multimarket interactions, and to measuring the impact of price stabilization on other variables of interest to governments other than measures of the producers' and consumers' surpluses. Our main contribution in this respect has been to show how agricultural sector models of the kind discussed in Chapter 4 can be used to evaluate price stabilization.

NOTES

1. Newbery (1976) and Newbery and Stiglitz (1981) have given an example in which risk-averse behavior can lead to society being made worse off through costless price stabilization. The short-run Marshallian surplus analysis does not necessarily imply risk neutrality (risk costs can be deducted as in Chapter 4), but it is implied when price stabilization is evaluated without allowing for any changes in the risk costs charged to producers as part of the area under the supply schedule.

2. Introduction of a multiplicative risk term in demand does not change our basic results as long as the supply and demand risks are assumed to be independently distributed (Scandizzo, Hazell, and Anderson, 1983).

3. Since the model is nonlinear, $\bar{p}$ will typically be different from the expected market clearing price in the prestabilized market.

4. Jensen's inequality states that, for any continuous and twice differentiable function $f(x)$ of a random variable x, $E[f(x)] - f(E[x]) \lessgtr 0$ as $f''(x) \lessgtr 0$. When applied to an integral function such as

$$f(x) = \int_{x}^{E[x]} h(x)dx$$

then

$$f(E[x]) = \int_{E[x]}^{E[x]} h(x)dx = 0,$$

and the inequality reduces to $E[f(x)] \lessgtr 0$ as $f''(x) \lessgtr 0$.

5. Since the observations generated in the simulation are free of statistical errors, we used the adjusted R^2 simply as a measure of accuracy in the choice of functional form. Obviously, unadjusted R^2s of 1.0 could be obtained with high order polynomials, but the adjusted R^2 will typically peak at lower values as higher order terms add coefficients faster than they increase the unadjusted R^2. We found that quadratic equations performed well by the adjusted R^2 criterion and, in more than half the cases, obtained adjusted R^2s of 0.7 or higher with sample sizes of 27. The revenue expectations models performed least well, with five of the six adjusted R^2s falling in the range of 0.34 to 0.48. The only real problem arose in the regression of producers' gains when $P_t^* = \bar{P}$. The adjusted R^2 was -0.05 for the quadratic model. The results for this equation are reported in Table 5.2 only for completeness.

6. Newbery and Stiglitz (1981) represent an important exception.

7. We assume c_x, c_e, and c_m are not stochastic; hence the variance of income and total revenue are identical.

8. Let $\tilde{p}_j = p_j + k$ with k constant, then $\text{cov}(\tilde{p}_j n_j, p_i n_i) = \text{cov}(p_j n_j, p_i n_i) + k\,\text{cov}(n_j, p_i n_i)$.

9. Complications arise when a stabilized commodity is traded internationally at uncertain prices. The agency must then have the financial facilities to stabilize these prices for the domestic market. Fortunately this problem does not arise with beans in Guatemala because they are not traded.

10. Note that the linearization is especially simple for the Guatemalan model since the B matrix is specified as diagonal.

11. The middle-sized farms have some idle land and labor resources in the model solution (which accords well with available evidence), and these resources are utilized more fully in the model when bean prices are stabilized.

6
Project Evaluation
in Risky Markets

Project evaluation is set typically in a deterministic model of an economy where, presumably for reasons
of ease and expediency, risk is assumed away as being
irrelevant or unimportant. Such deterministic analysis
can be shown to be appropriate even where risk is preva-
lent, provided that random components of supply and
demand schedules are specified as being additive in
nature so that only the intercepts of the supply and
demand curves are stochastic (Anderson 1983). However,
life is not quite so simple when risk enters into production in a multiplicative way.

Risk in agricultural production carries four basic
consequences for a market economy. First, physical output is contingent on the state of nature that happens to
be realized at any particular moment. Second, partly as
a consequence of the randomness of yields and partly due
to the operators' attempts at making production plans
based on forecasts, prices are stochastic too. The com-
plexity of this process, as noted in Chapter 2, is
increased significantly if risk enters multiplicatively
rather than additively. Third, attitudes towards risk,
if other than neutral, are important in determining production levels. Fourth, trade performance, market
clearing, and balance of payments equilibria can only be
judged by suitable statistics, and do not hold
invariantly except in a trivial ex post sense.

A feature common to all these consequences is the
'wedge' created by risk between operators' production
and consumption plans, and the realization of these
plans. For all economic variables, two separate
accounts can be formed -- namely ex ante and ex post
accounts or, alternatively, 'expectations' and 'realiza-
tions' accounts. We explored such risk accounting in
Chapter 2 and from slightly different perspectives in
Chapters 4 and 5. In this chapter we seek to relate
some of our earlier results to the topic of appraisal of

development projects. This necessarily involves us in contemplation of issues in shadow pricing.

A PARTIAL EQUILIBRIUM APPROACH TO SHADOW PRICES AND FORECASTS

Consider a traded good. If the international supply of this good is risky so is its price. The opportunity cost to the country of an extra unit of production of this good in a marginal project is the international price which is not known with certainty when the project is appraised. A suitable forecast is necessary and it is thus an ex ante shadow price.

It would seem intuitively true that an unbiased forecast of the international price should be the shadow price to be used. If such a forecast was available, ex ante and ex post shadow prices would coincide on average and the consequences of forecast error would be minimized. However, as should be expected from Chapter 2, the unbiased forecast will be inadequate if the project's output is stochastically correlated with the international price.

The inadequacy of an unbiased forecast for valuing the opportunity cost of risky production can be seen by focusing attention on the definition of the variables that are the object of the forecast. If the outcome of production is known, the problem of shadow pricing is what value should be placed on an additional unit of production. Under risk, this additional unit can be specified only conditional on a specific state of nature. Thus if in state i, production is Q_i and realized price is P_i, a meaningful full-information forecast will concern $\partial(P_i Q_i)/\partial E[Q_i] = r_i$ which is the (average) value obtained for an average additional unit of production. If prices or yields are either non-stochastic or noncorrelated, no distortion will arise but, if they are jointly random, an unbiased forecast of unit revenue rather than price is in order.

While the fact that an unbiased forecast of unit revenue rather than price is needed may be revealing, it does not readily suggest a class of forecasts to be used. Moreover, unit revenue is a rather difficult concept to specify and measure and market information usually comes in the form of realized sales and prices. Since a production plan will generally consist of a set of anticipated production levels for which there are no previous observations, it is thus still proper to ask what type of forecast would be best to appraise alternative production levels. From our earlier considerations we know that such a forecast should not be unbiased; what then would be the optimal distortion?

The easiest way of obtaining some specific rules of

thumb for calculating such a distortion is to make some specific but simple assumptions about the market structure. Suppose, for example, that the economy is closed, and that linear versions of Equations (2.1) and (2.3) provide reasonable approximations to the market structure. An appropriate model is then:

(6.1) $$S_t = \lambda \epsilon_t P_t^*,$$

(6.2) $$D_t = a - bP_t,$$

(6.3) $$S_t = D_t$$

and we again assume that $E[\epsilon_t] = \mu$, $V[\epsilon_t] = \sigma^2$, cov $[\epsilon_t, P_t^*] = 0$ for all t, where P_t^* is the price anticipated by producers at the time of making production decisions, ϵ_t is stochastic yield, and a, b, and λ are positive constants.

Assuming that the market clears in each period, market price is

(6.4) $$P_t = a/b - (\lambda/b)\, \epsilon_t\, P_t^*.$$

Taking expectations over ϵ_t and P_t^*, expected market price is

6.5) $$E[P_t] = a/b - (\lambda/b)\mu E[P_t^*].$$

Now if P_t^* is formed in such a way that $\lim_t E[P_t^*] = \lim_t E[P_t]$, that is, the forecast is asymtotically unbiased, then $E[P_t]$ will converge to a limiting value of

(6.6) $$E[P] = \lim_t E[P_t] = a/(b + \lambda\mu).$$

Consider now the shadow pricing question. On the demand side, as in earlier chapters, a measure of value can be defined as consumers' willingness to pay as measured by the area under the inverse demand curve

(6.7) $$W_{ct} = \int_0^{D_t} (a/b - D/b)\, dD.$$

Solving the integral in (6.7),

(6.8) $\qquad W_{ct} = (a/b)D_t - D_t^2/(2b).$

Since the market clears in every period, $D_t = S_t = \lambda \epsilon_t P_t^*$, thus yielding

(6.9) $\qquad W_{ct} = (a/b) (\lambda \epsilon_t P_t^*) - (\lambda \epsilon_t P_t^*)^2/(2b).$

Taking expectations over ϵ_t and P_t^* yields

(6.10) $\qquad E[W_{ct}] = (a/b)(\lambda \mu E[P_t^*]) - \lambda^2 \mu_2 E[P_t^{*2}]/2b.$

where μ_2 is the second moment about the origin of ϵ. Defining $C[\epsilon]^2 = \sigma^2/\mu^2$, the square of the coefficient of variation of yield, we can rewrite Equation (6.10) as

(6.11) $\qquad E[W_{ct}] = (a/b) E[S_t] - \{(1 + C[\epsilon]^2) E[S_t]^2$

$\qquad\qquad\qquad - (1 + C[\epsilon]^2) V[S_t]\} /2b.$

A shadow price, $\tilde{P}_s$ is now defined as the value that an average additional unit of expected supply has for the consumers

(6.12) $\qquad \tilde{P}_s = \partial E[W_{ct}]/\partial E[S_t]$

$\qquad\qquad = a/b - (1 + C[\epsilon]^2)E[S_t]/b.$

In order to find what we will term the first-best shadow price, $\tilde{P}_s$, we assume that all producers' forecasts converge in mean and that this mean equals $\tilde{P}_s$. Thus $E[P^*] = \tilde{P}_s$ and

(6.13) $\qquad \tilde{P}_s = a/\{b + \lambda \mu (1 + C[\epsilon]^2)\}.$

In order to find the relative magnitude of the distortion, divide $\tilde{P}_s$ by $E[P]$ as given in (6.6),

(6.14) $\qquad \tilde{P}_s/E[P] = (b + \lambda \mu)/\{b + \lambda \mu (1 + C[\epsilon]^2)\}.$

That this ratio is less than unity is apparent by inspection but can also be seen as follows. Denoting by $|\xi| = b/(\lambda\mu)$, the absolute value of the ratio between the demand and supply elasticities evaluated at the expected market price $E[P]$, we can show that

$$(6.15) \qquad \tilde{P}_s/E[P] = (1 + |\xi|)/(1 + C[\varepsilon]^2 + |\xi|) < 1.$$

We believe that analogous results would in principle be obtained in more general equilibrium models of a closed economy such as described in Chapter 2--although they would not have the simple and immediately applicable form of our suggested rule of thumb which can be applied easily even when relevant time series of revenues are absent. For an open economy, such results would not be found unless one takes the welfare of all countries participating in a market as the norm against which to measure project benefits, rather than the welfare of the particular country in which a project happens to be located. For the so called 'weak commodities', UNCTAD has repeatedly suggested that such a commodity orientation in project evaluation may be in order to take into account the negative externalities generated for all present producers by a new project.
Our rule of thumb represented by Equation (6.15) tells us that projects for the risky production of commodities with demand elasticity low relative to supply elasticity should be appraised at shadow prices below expected market prices. Such appraisal will tend to constrain the production of these commodities and will thus influence the distribution of welfare between developing and industrialized countries and the total welfare of consumers in all countries.

PROJECT EVALUATION AND INCOME DISTRIBUTION

The General Problem of Project Evaluation Under Risk

A more rigorous and general treatment of risk and income distribution in project analysis cannot be avoided. The treatment of risk in project analysis has generally been confined to procedures to quantify the effects on the decision criterion (for example, the internal rate of discount) of specific risky outcomes (Reutlinger 1970, Pouliquen 1970). Most events contemplated in the life of a project are affected by various risks and both the benefits and costs are typically contingent on risky states of nature. The established techniques for appraisal under uncertainty concern the use of appropriate methods of prediction, allowances for risk and risk aversion and various forms of stochastic sensitivity analysis (Anderson, Dillon and Hardaker 1977, chap. 8).

These methods, however, useful as they can be when Pareto-optimal equilibria can be reached without need of government interventions, do not address the problem of risk when it results in market failures through the disruption of the orderly behavior of economic agents and the introduction of difficulties and errors in investment and production decisions. In such cases market prices cannot be considered as adequate signals for project selection and neither averaging or risk allowance techniques, per se, can help in improving the allocation of resources.

Furthermore, the established techniques do not address the case where a social welfare function (or system of distribution weights) is used to rank income distribution across economic agents, thereby ignoring the potentially important implications for ranking risky prospects involving agents' diminishing marginal utility of income.

Because income distribution considerations are of uncontroversial importance in developing countries and because of the unexplored relationship between individuals' risk aversion and egalitarian preferences, in this section we attempt a recasting of the partial equilibrium results for shadow prices under risk within the general framework of an expected social welfare function. To proceed to such a task, we first review the general methodology of a project evaluation based on (a) income distribution weights for riskless events and on (b) Pareto-better risky projects. The problems of integrating equity and risk considerations are then addressed.

Some Specific Considerations of Welfare Changes in Project Evaluation

Consider the general form of a social welfare function

$$(6.18) \qquad W = W(U_1, U_2, \ldots U_I, U_g),$$

where W is an ordinal indicator of social welfare, and U_i, U_g are ordinal indicators of utility for the ith consumer, $i = 1, 2, \ldots, I$, and for the government, g, respectively. Each of these utility indicators is, in turn, a function of the quantities consumed of a total of J goods, as in

(6.19) $$U_i = U_i (c_{i1}, c_{i2}, \dots, c_{iJ});$$

$$i = 1, 2, \dots, I, g$$

where c_{ij} represents the quantity of good j consumed by consumer i.

The effect of a project will be to alter the configuration of the economy as measured by the levels of total and individual utilities so that, for a marginal increment in welfare dW,

(6.20) $$dW = \Sigma_i \Sigma_j (\partial W/\partial U_i)/(\partial U_i/\partial c_{ij})dc_{ij}$$

$$+ \Sigma_j (\partial W/\partial U_g)(\partial U_g/\partial c_{gj})dc_{gj}.$$

Assuming market equilibrium and no distortions, the first-order conditions of individual utility maximization (cf. Equation (2.34)) give the result that

(6.21) $$\partial U_i/\partial c_{ij} = (\partial U_i/\partial Y_i)P_j,$$

$$i = 1, 1, \dots, I, g;$$

$$j = 1, 2, \dots, J$$

where P_j is the price of the jth good.

Substituting (6.21) into (6.20) yields

(6.22) $$dW = \Sigma_i (\partial W/\partial U_i)(\partial U_i/\partial Y_i)dY_i$$

$$+ (\partial W/\partial U_g)(\partial U_g/\partial Y_g)dY_g,$$

where $dY_i = \Sigma_j P_j dc_{ij}$, $i = 1, 2, \dots, I, g$.

Because W is ordinal we can divide both sides by the marginal social welfare of government income without changing the underlying system of social preferences. Thus

(6.23) $$dW = dW/(\partial W/\partial Y_g)$$

$$= dY_g + \Sigma_i [(\partial W/\partial U_i)(\partial U_i/\partial Y_i)/(\partial W/\partial Y_g)]dY_i.$$

Using the notation $w_i = (\partial W/\partial Y_i)/(\partial W/\partial Y_g) = \partial Y_g/\partial Y_i$ and adding and subtracting $\Sigma_i dY_i$ from (6.23) we obtain

(6.24) $$dW = dY + \Sigma_i (w_i - 1)dY_i,$$

where $dY = \Sigma_i dY_i + dY_g$. From (6.24), a necessary condition for the 'efficiency gain', dY, to measure correctly the benefits of a project is that

$$(6.25) \qquad \Sigma_i (w_i - 1)dY_i = 0.$$

If, as is typically the case, the dY_i are all non-negative, and if the w_i are all less than unity (that is, private consumption generated by the project is considered to be less socially valuable than income generated for the government), then Equation (6.25) cannot hold true and the efficiency gains taken above will overstate the societal gain from the project. Alternatively, if the acceptance criterion used in project evaluation is $dY \geq 0$, then projects will be accepted even though dW may be negative. Thus, relative to such an efficiency-oriented analysis in which the w_i are considered, social analysis makes project selection that much more conservative. To sum up from the results in (6.23) through (6.25), a set of sufficient conditions for the efficiency criterion to coincide with the social criterion is that first, society is indifferent as to the utility gained by different consumers ($\partial W/\partial U_i$ is constant) and second, the marginal utilities of all consumers and the government are identical.

Alternative Approaches to the Estimates of Social Weights

Equation (6.24) can also be written as

$$(6.26) \qquad dW = dY + \Sigma_i ((\partial W/\partial Y_i)/(\partial W/\partial Y_g) - 1)dY_i.$$

Several approaches to estimate the relative weights $(\partial W/\partial Y_i)/(\partial W/\partial Y_g) = \partial Y_g/\partial Y_i$ are possible. First, without loss of generality, we can write (6.26) as

$$(6.27) \qquad dW = dY + \Sigma_i ((\partial U_i/\partial Y_i)/(\partial U_g/\partial Y_g) - 1)dY_i,$$

which is an expanded version of (6.24).

If we introduce the conventional simplifying assumption that all consumers possess identical utility functions, the weighting system in (6.27) will depend primarily on their income levels. For example, assuming that $\partial U_g/\partial Y_g = U_g'$ is known and that

$$(6.28) \qquad \partial U_i/\partial Y_i = Y_i^{-n} \ , \ i = 1, \ 2, \ \ldots \ , \ I,$$

we obtain a specific version of (6.27) as

$$(6.29) \qquad dW = dY + \Sigma_i (Y_i^{-n}/U_g' - 1) \ dY_i.$$

We can thus identify a weighting system $w_i = Y_i^{-n}/U_g'$ depending on a 'private parameter' n, the assumed constant elasticity of marginal utility, and a social parameter U_g'. Several approaches are available for determining the schedule of the private marginal utility of income such as the Frisch-parameter approach proposed by Balassa (1977) or the direct measurement of consumers' preferences.

An alternative approach is to assume that a marginal utility schedule of the type given in (6.28) may be taken to subsume both public and private preferences, as in

$$(6.30) \qquad \partial W/\partial Y_i = Y_i^{-n} \ , \ i = 1, \ 2, \ \ldots \ , \ I.$$

The parameter n can be decomposed into 'public' and 'private' components. Let π represent the elasticity of private marginal utility of income and let α denote the proportional increase in marginal social welfare caused by a small proportional increase in the marginal utility of the representative consumer. Then $n = \pi\alpha$ and

$$(6.31) \qquad \partial W/\partial Y_i = (Y_i^{-\pi})^{\alpha}.$$

Project Evaluation Under Risk

In principle, the incorporation of risk in the welfare framework of project evaluation can be done in a way similar to that just described for income distribution effects. In both cases the role performed by the welfare function is to rank income distributions -- in the income case, over different consumers and in the risk case over states of nature.

Assuming that risk enters production in a multiplicative fashion, we can postulate the following equality for a closed economy

(6.32) $\qquad c_j = q_j^* \varepsilon_j / \mu_j \;, \; j = 1, \; 2, \; \ldots \;, \; J \;,$

where c_j is the quantity consumed of the jth good, q_j^* is the anticipated production index, and ε_j is a random variable with mean μ_j and variance σ_j^2. Without losing all generality, we can abstract from the income distribution problem by assuming that c_j denotes the consumption of a representative consumer in a perfectly egalitarian society.

Considering a welfare function as a reduced form of (6.18) and (6.19),

(6.33) $\qquad W = W(c_1, \; c_2, \; \ldots \;, \; c_J),$

the effect of a marginal project on the expected value of W is

(6.34) $\qquad dE[W] = E[\Sigma_j \; (\partial W / \partial c_j)(\varepsilon_j / \mu_j) d(q_j^*)]$

and, assuming again market equilibrium with no distortions,

(6.35) $\qquad dE[W] = E[\Sigma_j \; (\partial W / \partial Y)(\varepsilon_j / \mu_j) P_j d(q_j^*)].$

Standardizing this with the marginal social welfare W', as in (6.23), and factoring out the jth marginal effect gives the jth shadow price

(6.36) $\qquad \tilde{P}_j = \partial (E[W] / W') / \partial (q_j^*)$

$\qquad\qquad = E[((\partial W / \partial Y) / W')(\varepsilon_j / \mu_j) P_j],$

where W' is the marginal utility of income evaluated appropriately, such as at the average revenue $P_j q_j^*$. For a welfare function with constant marginal utility of income, the shadow price simplifies to average revenue $E[\varepsilon_j P_j] / \mu_j$. Equation (6.36) can be expressed alternatively as

$$(6.37) \qquad \tilde{P}_j = E[\varepsilon_j P_j]/\mu_j + E[((\partial W/\partial Y)/W' - 1)\varepsilon_j P_j/\mu_j]$$

and, similarly, Equation (6.35) as

$$(6.38) \qquad dE[W^*] = dE[W]/W'$$

$$= dE[Y] + E[\Sigma_j((\partial W/\partial Y)/W' - 1)P_j$$

$$(\varepsilon_j - \mu_j)/\mu_j d(q_j^*)].$$

Approximating $\partial W/\partial Y$ with a Taylor expression truncated after the first-order term, and substituting into (6.38) we obtain

$$(6.39) \qquad dE[W^*] = dE[Y]$$

$$-\phi \sum_{j=1}^{J} \sum_{k=1}^{J} \frac{cov[P_j\varepsilon_j,\ P_k\varepsilon_k]}{\mu_j\mu_k} d(q_j^*),$$

where ϕ is the coefficient of absolute risk aversion[1] evaluated at $E[Y]$.

In expressions (6.38) and (6.39), we have assumed that the effect of a project is to shift only the average values of revenues and incomes while leaving the distributions of the underlying random variables unaltered. This need not be the case as many projects will change the amount of risk. The assumption is made for the sake of simplicity as a change in risk due to the project is really a separate component of the benefits (costs) of the project.

Accounting for Both Equity and Risk

The most direct way to include both equity and risk considerations in the calculation of a project's benefits is to take the expected value of expression (6.24)

$$(6.40) \qquad dE[W] = dE[Y] + E[\Sigma_i(w_i - 1)dY_i].$$

A specific form of the social welfare function which assigns curvature characteristics over income groups will now have these characteristics imparted over the states of nature. In other words, for any selected set of income distribution weights, the project evaluator or the policy maker will be committed to a given degree of risk aversion, and vice versa.

While this direct approach may be simple, its implications are worrying. First, it is easy to imagine cases where value judgments about the desirability of better income distributions would imply severe degrees of risk aversion. Second, empirical measurements and theoretical considerations suggest different ranges of the values for income distribution and risk aversion parameters.[2] Third, while risk aversion appears to be more reasonably embedded in individual behavior, egalitarian preferences are not directly reflected in individual choices and are more of an ethical nature.

Rather than lumping these effects together, it appears desirable to separate them by decomposing social welfare into two distinct expectations, one taken with respect to states of nature and one taken over individuals. More specifically, given a welfare function of the form $W = W(Y_1, Y_2, \ldots Y_I)$, we can decompose its expectations using the mean value theorem as

$$(6.41) \qquad E[W] = W(\overline{Y}_1, \overline{Y}_2, \ldots, \overline{Y}_I)$$
$$+ E[\Sigma_i W_i (Y_i - \overline{Y}_i)]$$

where $\overline{Y}_i = E[Y_i]$ and $W_i = \partial W / \partial Y_i$ evaluated at a point $\tilde{Y}_i$ between Y_i and $\overline{Y}_i$, $\tilde{Y}_i = Y_i + \alpha(\overline{Y}_i - Y_i)$, $0 \leq \alpha \leq 1$), $i = 1, 2, \ldots, I$.

In Equation (6.41) the first term represents the ranking of consumers 'riskless' incomes, while the second term is a social 'risk premium.' That is, it represents the excess social welfare required over and above the welfare provided by an equivalent and riskless constellation of average incomes to compensate for the random variations in individual consumption.

Equation (6.41) demonstrates that it is not possible to make separate judgments about the income distribution and risk aversion utility parameters within a straightforward expected utility framework. Once a functional form is selected for the riskless ranking of average incomes in the first right-hand term of (6.41), one is committed to a particular functional form for the marginal utility and, therefore, for the risk premium. The reverse is also true. Once the risk premia are chosen for each income class, the functional form of the welfare function is also determined up to an additive constant.

The implications of this inseparability for project appraisal can be analyzed by considering the increase in social welfare induced by a marginal investment as in

(6.42) $\quad dE[W] = \Sigma_i W_i d\bar{Y}_i + E[\Sigma_i W_i (dY_i - d\bar{Y}_i)]$

$$+ E[\Sigma_i W_{ii} (Y_i - \bar{Y}_i) dY_i],$$

where $W_{ii} = \partial^2 W/\partial Y_i^2$.

If we evaluate the first and second derivatives at $Y_i = \bar{Y}_i$, so that $\bar{W}_i = \partial W/\partial Y_i \mid \bar{Y}_i$ and $\bar{W}_{ii} = \partial^2 W/\partial Y_i^2 \mid \bar{Y}_i$, we can write (6.42) in the more transparent form

(6.43) $\quad dE[W]/W_g = \Sigma_i \bar{W}_i/W_g - d\bar{Y}_i + \Sigma_i (\bar{W}_{ii}/W_g)$

$$\bar{Y}_i C [Y_i]^2 d\bar{Y}_i$$

where W_g denotes the marginal expected welfare of government income and $C[Y_i]$ is the coefficient of variation of Y_i.[3] Noting that the second term of (6.43) contains measures of local curvature of the welfare function, $-\theta_i = \bar{W}_{ii}\bar{Y}_i/W_g$, that are unit free and similar to a coefficient of relative risk aversion, Equation (6.43) can be rewritten as

(6.44) $\quad dE[W] = \Sigma_i d\bar{Y}_i + \Sigma_i (\bar{w}_i - 1) d\bar{Y}_i$

$$- \Sigma_i \theta_i C [Y_i]^2 d\bar{Y}_i$$

or

$$\begin{pmatrix} \text{Increase} \\ \text{in} \\ \text{welfare} \end{pmatrix} = \begin{pmatrix} \text{Increase} \\ \text{in average} \\ \text{income} \end{pmatrix} + \begin{pmatrix} \text{Income} \\ \text{distribution} \\ \text{effect} \end{pmatrix}$$

$$- \begin{pmatrix} \text{Risk} \\ \text{premium} \end{pmatrix}$$

Using the relationship

(6.45) $\quad dY_i = \Sigma_j \xi_j^{(i)} (P_j \varepsilon_j/\mu) d(q_j^*),$

where $\xi_j^{(i)}$ is the ratio between the own-price demand and supply elasticities for the ith consumer, we obtain

(6.46)
$$dE[W] = \Sigma_i d\bar{Y}_i + \Sigma_i(w_i - 1)d\bar{Y}_i$$

$$- E[\sum_{i=1}^{I} \sum_{j=1}^{J} \sum_{k=1}^{J} \phi_i \xi_j^{(i)}$$

$$\frac{cov(P_j\varepsilon_j \ P_k\varepsilon_k)}{\mu_j\mu_k} \ d(q_j^*)]$$

where $\phi_i = \bar{W}_{ii}/W_g$. Expression (6.46) mirrors Expression (6.39) and can be used to obtain the shadow price $\tilde{P}_j$

(6.47)
$$\tilde{P}_j = \partial E[W]/\partial(q_j^*)$$

$$= E[\varepsilon_j P_j/\mu_j] + E[\Sigma_i \xi_j^{(i)}(w_i - 1)\varepsilon_j P_j/\mu_j]$$

$$- 2E[\sum_{k=1}^{J} \sum_{i=1}^{I} \phi_i \xi_j^{(i)}](cov[P_j\varepsilon_j P_k\varepsilon_k]/\mu_j\mu_k).$$

Alternative Approaches to the Estimation of Risk Aversion

While several direct ways may be used to estimate the risk premia (see for example, Anderson, Dillon and Hardaker 1977, chap. 4), it seems reasonable to assume that in project evaluation risk premia should be set to reflect the risks and preferences of the decision makers involved. From the point of view of the economist appraising a rural development project, a risk premium can be conceived as the compensation required by the farmers over and above the money value of the project to be indifferent between the risky project and an equally valuable certain alternative.

In this case, risk premia can be measured by direct elicitation from the farmer, or through indirect examination of farm records and statistics. In the first case, they are estimated by asking the farmers hypothetical questions to determine certainty equivalents of particular prospects.

In evaluating risky alternatives for subsistence farmers, one has also to consider carefully the potential impact of the risk of a serious disaster. Casual observations and some empirical studies reveal particularly cautious behavior of farmers when there is a possibility of bankruptcy or of a loss of subsistence consumption. Under these circumstances, the concept of a simple risk premium may be questionable.

An aggregate alternative to private elicitation of risk premia is provided by the iso-elastic utility function. Using the form $W = Y^{1-n}/(1 - n)$, where $Y = \Sigma_i Y_i$, and applying expression (6.44), we obtain

$$(6.48) \qquad dE[W] \doteq (\bar{Y}/Y_g)^{-n} \Sigma_i \ (1 - nC[Y_i]^2)d\bar{Y}_i$$

where $\bar{Y}$ equals $\Sigma_i \bar{Y}_i$ and government income Y_g is taken to be riskless. For each income group, the income distribution weight under risk will thus equal the riskless weight minus a risk premium depending on the coefficient of variation and on the utility parameter n. Some weights are exemplified in Table 6.1.

The expressions obtained for the income distribution weights under risk, and especially Equation (6.48) above, have two main implications for practical project evaluation. First, the values of n associated with judgments about income distributions should account for the fact that part of the weight given to each income class to reflect its income distribution position is counterbalanced by the risk premium associated with each income increase. The distribution of the risk premia will depend on whether the coefficient of variation of

TABLE 6.1
Riskless and Risky Distribution Weights[a]

Per Caput Income $	$C[Y_i]$	Riskless Distribution Weight (n=2)	Risky Distribution Weight (n=2)
100	0.50	100.00	59.30
250	0.35	16.00	14.33
500	0.32	4.00	3.77
750	0.30	1.77	1.72
1,000	0.28	1.00	1.00
1,500	0.26	0.44	0.45
3,000	0.24	0.11	0.12
6,000	0.20	0.03	0.03
10,000	0.10	0.01	0.01

[a] Normalized in terms of an average consumption level of $1,000 per caput.

income increases or decreases with the size of income. However, if the coefficient of variation is constant across income classes, and for the government, no modifications will be caused to the system of income distribution weights as a result of considering risk. If poor people have riskier incomes, as it is often the case, the impact of the risk premium will make the system of riskless weights less progressive, particularly at the extremes.

Second, the judgment on the risk premia will depend on the riskiness of the project only through the estimate of $d\bar{Y}_i$, unless project risks are such that they cause a detectable increase or decrease in the overall coefficient of variation of the beneficiaries' incomes.

Shadow Pricing Under Risk

Expression (6.48) can be used directly to find approximate values for shadow prices of each project good. In general, such shadow prices will not be needed if risk modified-social weights are used to appraise the project because in such cases it is sufficient to apply the weights to the expected incremental incomes generated by the project. Nevertheless, it is useful to compare the implicit shadow prices generated by the distribution weights with the shadow prices suggested in our partial equilibrium analysis.

For the case of a producer good, if we define income according to

$$(6.49) \qquad Y_i = \Sigma_j (P_j \varepsilon_j^{(i)}/\mu_j) q_j^{*(i)}$$

where $q^{*(i)}$ is the average quantity produced of good j by income group i and $\varepsilon_j^{(i)}$ is a group-specific random disturbance. An income-group specific shadow price can be derived directly from (6.48) as

$$(6.50) \qquad \tilde{P}_j^{(i)} = \frac{dE[W]/dq_j^{*(i)}}{dE[W]/dY\big|_{\bar{Y}}} \{1 - C[Y_i]^2\}$$

$$E[P_j \varepsilon_j^{(i)}/\mu_j].$$

Note that, for each producer, the relative risk premium is independent of the particular good considered. Thus the relative shadow price of the ith producer with respect to any of his products taken as a numeraire is

still his own revenue expectation. Second, two elements
of risk appear to be present in each project: (a) a
general risk associated with the beneficiary's income,
and (b) a product-specific risk associated with the risk
in the production of the jth (project) good. As already
noted, if expected welfare weights are used to appraise
the project, the first-best price is the revenue expectation.

If revenue expectations are available, therefore,
their use in project evaluation as shadow prices will be
all that is needed provided that an appropriate system
of income distribution weights is used. As a shortcut,
when revenue-based forecasts are not available to the
project analyst, the rules of thumb obtained through
partial equilibrium analysis can be used along the same
lines. Finally, if distribution weights are used,
expression (6.50) can provide a product-by-product
shadow price.

CONCLUSIONS

The analysis of this chapter maintains and reinforces the basic prescription that revenue-based forecasts should be used to guide productive decisions under
risk. However, since forecasts should be made by decision agents on the basis of their own marginal revenues,
one may anticipate a scarcity of revenue statistics for
new production enterprises which are typically the
object of project evaluation. In these cases, some
assessment of the market demand elasticity and of the
coefficient of variation of production would be sufficient to improve substantially naive forecasts based
on expected values of prices and yields.

When the purpose of project evaluation is considered
to be the maximization of a social welfare function, income distribution and risk aversion considerations come
into play. Within the usual framework of expected utility, these considerations are not separable: a given
system of income distribution weights implies a corresponding degree of social risk aversion. Conversely a
given risk premium implies a specific ranking of consumers' incomes.

If these interdependencies are taken into account
and social parameters are specified on the basis of
their implications for income distribution and risk
aversion, unbiased revenue forecasts appear to be the
appropriate shadow prices to be used in project evaluation. When social analysis is not carried out, on
the other hand, revenue forecasts are still an appropriate second-best statistic, provided that they are
correctly estimated for each group of project beneficiaries.

NOTES

1. The coefficient of absolute risk aversion is defined as the ratio $-(\partial^2 W/\partial Y^2)/(\partial W/\partial Y)$.

2. See, for example, Little and Mirrlees (1974), Moscardi and de Janvry (1978), Dillon and Scandizzo (1978), Hamal and Anderson (1982), and Anderson (1983).

3. Based on the multiplicative form $Y_i = \bar{Y}_i \varepsilon_i$, where ε_i is a multiplicative disturbance with mean μ and variance σ^2.

7
Epilogue

We have now considered a variety of topics related
to the behavior of economic agents, especially producers, in risky markets. Our style has been to simplify
our models of these markets to the important essence of
reality as we see it -- namely producers who are averse
to risk, facing a risky productive environment yet being
obliged to commit resources before they know what chance
will deliver them in yields and prices. Producers must
then form anticipations about prices and yields, thereby
raising the prospect of losses in market and resource
allocative efficiency that may be costly to their own
income and to consumers' welfare.

Our central finding is that in competitive markets
of the type specified, producers should take expected
per unit revenue as their rational price forecast so
that they properly account for covariance effects between their individual production and realized prices.
The expected per unit revenue forecast maximizes the
average value of producers' surplus. It is also the
price forecast that ensures the largest value of social
welfare as measured by the sum of the expected producers' and consumers' surpluses.

If prices and yields are negatively correlated, the
expected per unit revenue will be less than the average
market clearing price. In this case, producers should
produce less of the commodity in question than calcula-
tions based on average market price would suggest; a
point often overlooked by many economists and policy
makers. The opposite will happen when the correlation
is positive. These supply effects will arise even if
producers are risk neutral.

In their observed behavior it is not easy to see how
producers really do form their anticipations. However,
in our several attempts at supply analysis with contrasting arguments we concluded that many farmers seemingly hold their anticipations in an appropriate way.
Those that do not tend to be found in either developing

123

countries or in centrally planned economies. There may be good reasons why such joint price-yield effects either can or must be ignored in these countries, but our findings imply that these are the cases where there is relatively greater scope for policy intervention to improve market efficiency.

Two types of losses in producers' and societal welfare arising from forecasting errors can usefully be distinguished. The first type of loss arises when producers fail to plan on the basis of expected per unit revenues. In this case the forecast errors are larger than they need be. The second type of loss arises even when producers are rational in their expectations; it is the part of the forecast error that can only be removed by stabilizing prices and yields. Our numerical analyses show that the second type of loss is relatively small, but it increases with the coefficient of variation of yields and with the inelasticity of demand, and it is smaller the more risk averse are producers.

The first type of loss can be much larger, particularly if producers hold such naive price expectations as previous-period price. However, the intersection price between demand and expected supply can be almost as efficient as the expected per unit revenue forecast.

When markets are very inefficient, a government can choose between three basic policy approaches. First, production quotas could be imposed to limit the average output levels of risky crops to their socially desired norms. Second, market information services could be established or improved to help producers forecast better. Third, the government could attempt to reduce or eliminate risks from the market through price stabilization, futures markets or crop insurance.

We have explored the price stabilization option with the aid of simulation and mathematical programming models. Our findings show that the social return from price stabilization is likely to be much less than the cost of achieving it where producers initially plan on the basis of expected per unit revenue. Larger gains are possible where less appropriate price forecasting is pursued, but then the largest part of the gain might also be more effectively attained through improved market information services. These findings strongly reinforce the conclusions arising from theoretical work of Newbery and Stiglitz (1981).

We have not attempted to analyze futures markets or crop insurance in this study. Work by Cox (1976) and Peck (1976) amongst others suggests that futures markets may have a useful role to play in reducing the variability of forward prices for producers, though it is doubtful such markets are institutionally viable for most developing countries. The effectiveness of crop insurance in reducing risks for farmers and in improving market and resource allocative efficiency has recently been reviewed by Hazell, Pomareda and Valdés (forthcoming). Except for insurance against very specific types of risks (for example, hail damage), it would seem that crop insurance as a general risk management approach is expensive, and has only proved viable when heavily subsidized by government.

We conclude that, for many countries, and particularly those in the developing world, the most effective way to improve the efficiency of risky markets is for government to provide an adequate program of data collection, appropriate forecasting, and information dissemination to assist farmers in forming their price and yield expectations. Such programs might range from assistance to farmers in recording and calculating weighted averages of past revenues and yields, to more elaborate market intelligence services that provide timely and detailed information throughout the year on prices, weather, sown areas and the like. Work by Hayami and Peterson (1972) and Freebairn (1976), and Smyth (1973) has shown how market information systems might usefully be evaluated in terms of their social costs and benefits. What is now needed is a much more thorough empirical understanding of the kinds of schemes that are useful and workable in developing countries.

Another upshot of our work is its implications for the specification of agricultural supply models in applied research. Since it is often difficult to tell how producers really form their price expectations, it seems that the appropriate way to specify supply functions for econometric estimation is to feature expected revenue (in lagged or subjective form) as the appropriate argument. Similarly, when it comes to building planning models for agricultural sectors (or indeed, for individual farms), it is again appropriate to attribute an ability to anticipate expected per unit revenues. This leads to a few complications in specifying appropriate mathematical programming models. However, with a little effort it can be done and, in elaborating the appropriate methods in Chapter 4, we suggest that, particularly when behavior is not very risk averse, it may be worth assuming such behavior routinely in sectoral models.

Finally, and perhaps unsurprisingly, we have argued that the idea of revenue expectations also has immediate relevance in techniques for evaluating investment projects. Appropriate shadow prices should embody the same effects that make expected prices differ from expected per unit revenues. More specifically, when risk aversion and income distribution considerations are taken into account through the use of a social welfare function, the shadow price of any project good should be the corresponding revenue expectation. Since time-series

References

Adams, F. G., and S. A. Klein. 1978. Stabilizing World Commodity Markets. Lexington: Lexington Books.

Anderson, J. R. 1974. "Risk Efficiency in the Interpretation of Agricultural Production Research." Review of Marketing and Agricultural Economics 42: 131-84.

_____. 1977. Methods and Programs for Analysis of Risky Gross Margins. Armidale: University of New England, Department of Agricultural Economics and Business Management, Miscellaneous Publications No. 5.

_____. 1983. "On Risk Deductions in Public Project Appraisal." Australian Journal of Agricultural Economics 27: 45-52.

Anderson, J. R., J. L. Dillon, and J. B. Hardaker. 1977. Agricultural Decision Analysis. Ames: Iowa State University Press.

Anderson, K. 1974. "Distributed Lags and Barley Acreage Response Analysis." Australian Journal of Agricultural Economics 18: 119-32.

Aoki, M. 1976. Optimal Control and System Theory in Dynamic Economic Analysis. Amsterdam: North-Holland.

Askari, H., and J. T. Cummings. 1976. Agricultural Supply Response: A Survey of the Econometric Evidence. New York: Praeger.

Atkinson, A. C. 1970. "A Method for Discriminating Between Models." Journal of the Royal Statistical Society 32, Series B: 323-45.

Balassa, B. 1977. "The Income Distribution Parameter in Project Appraisal." In Economic Progress, Private Values and Public Policy, B. Balassa and R. Nelson (eds). Amsterdam: North-Holland.

Behrman, J. R. 1968. Supply Response in Underdeveloped Agriculture. Amsterdam: North-Holland.

Bergendorff, H. G., P. B. R. Hazell, and P.L. Scandizzo. 1974. On the Equilibrium of a Competitive Market When Production is Risky. Washington, D.C.: World Bank, Development Research Center, mimeograph.

127

Blandford, D., and S. Lee. 1979. "Quantitative Evaluation of Stabilization Policies in International Commodity Markets." American Journal of Agricultural Economics 61: 128-34.

Brownlee, O. H., and W. Gainer. 1949. "Farmers' Price Anticipations and the Role of Uncertainty in Planning." Journal of Farm Economics 31: 266-75.

Campbell, R., B. Gardner, and H. Haszler. 1980. "On the Hidden Revenue Effects of Wool Price Stabilization in Australia: Initial Results." Australian Journal of Agricultural Economics 24: 1-15.

Castro, R. S., and J. A. Seagraves. 1974. The Supply of Winter Green Peppers in Florida. Raleigh: North Carolina State University, Economics Research Report No. 31.

Cochrane, W. W. 1980. "Some Nonconformist Thoughts on Welfare Economics and Commodity Stabilization Policy." American Journal of Agricultural Economics 62: 508-11.

Cox, C. C. 1976. "Futures Trading and Market Information." Journal of Political Economy 84: 1215-37.

Cox, D. R. 1961. "Test of Separate Families of Hypotheses." Proceedings of the Fourth Berkeley Symposium 1: 105-23.

_____. 1962. "Further Results on Tests of Separate Families of Hypotheses." Journal of the Royal Statistical Society 24, Series B: 406-24.

Dastoor, N. K. 1978. Non-nested Hypothesis Testing: The A. C. Atkinson Approach to Linear Regression Models. Mimeograph.

Day, R. H. 1963. "On Aggregating Linear Programming Models of Production." Journal of Farm Economics 45: 797-813.

Dillon, J. L., and P. L. Scandizzo. 1978. "Risk Attitudes of Subsistence Farmers in Northeast Brazil: A Sampling Approach." American Journal of Agricultural Economics 60: 425-35.

Duloy, J. H., and R. D. Norton. 1973. "CHAC, A Programming Model of Mexican Agriculture." In Multi-Level Planning: Case Studies in Mexico, L. Goreux and A. Manne (eds). Amsterdam: North-Holland.

Duloy, J. H., and R. D. Norton. 1975. "Prices and Incomes in Linear Programming Models." American Journal of Agricultural Economics 57: 591-600.

Feder, G. 1977. "The Impact of Uncertainty in a Class of Objective Functions." Journal of Economic Theory 16: 504-12.

Fisher, B. S., and R. G. Munro. 1983. "Supply Response in the Australian Extensive Livestock and Cropping Industries: A Study of Intentions and Expectations." Australian Journal of Agricultural Economics 27: 1-11.

Fisher, B. S., and C. Tanner. 1978. "The Formulation of Price Expectations: An Empirical Test of Theoretical Models." American Journal of Agricultural Economics 60: 245-48.

Fisher, R. A. 1920. "A Mathematical Examination of the Methods of Determining the Accuracy of an Observation by the Mean Error, and by the Mean Square Error." Royal Astronomical Society (monthly notes) 80: 758-69.

Fletcher, S. M., and C. Gellatly. 1977. On Econometric Analysis of the Supply Response for North Carolina Field Crops. Raleigh: North Carolina State University, Department of Economics and Business, mimeograph.

Freebairn, J. W. 1976. "Welfare Implications of More Accurate Rational Forecast Prices." Australian Journal of Agricultural Economics 20: 92-102.

Freebairn, J. W., and G. C. Rausser. 1975. "Effects of Changes in the Level of U.S. Beef Imports." American Journal of Agricultural Economics 57: 676-88.

Freund, R. J. 1956. "The Introduction of Risk into a Programming Model." Econometrica 24: 253-63.

Griffiths, W. E., and J. R. Anderson. 1978. "Specification of Agricultural Supply Functions -- Empirical Evidence on Wheat in Southern N. S. W." Australian Journal of Agricultural Economics 22: 115-28.

Hamal, K. B., and J. R. Anderson. 1982. "A Note on Decreasing Absolute Risk Aversion Among Farmers in Nepal." Australian Journal of Agricultural Economics 26: 220-25.

Hayami, Y., and W. Peterson. 1972. "Social Returns to Public Information Services: Statistical Reporting of U. S. Farm Commodities." The American Economic Review 62: 119-30.

Hazell, P. B. R. 1971a. "A Linear Alternative to Quadratic and Semivariance Programming for Farm Planning Under Uncertainty." American Journal of Agricultural Economics 53: 53-62.

_____. 1971b. "A Linear Alternative to Quadratic and Semivariance Programming for Farm Planning Under Uncertainty: Reply." American Journal of Agricultural Economics 53: 664-65.

Hazell, P. B. R., and C. Pomareda. 1981. "Evaluating Price Stabilization Schemes with Mathematical Programming." American Journal of Agricultural Economics 63: 550-56.

Hazell, P. B. R., C. Pomareda, and A. Valdés, (eds). Forthcoming. Agricultural Risks and Insurance: Issues and Policies. Baltimore: Johns Hopkins University Press.

130

Hazell, P. B. R., and P. L. Scandizzo. 1974. "Competitive Demand Structures Under Risk in Agricultural Linear Programming Models." American Journal of Agricultural Economics 56: 235-44.
_____. 1975. "Market Intervention Policies When Production is Risky." American Journal of Agricultural Economics 57: 641-49.
_____. 1977. "Farmers' Expectations, Risk Aversion and Market Equilibrium Under Risk." American Journal of Agricultural Economics 59: 204-09.
Hazell, P. B. R., R. D. Norton, M. Parthasarathy, and C. Pomareda. 1983. "The Importance of Risk in Agricultural Planning Models," In The Book of CHAC: Programming Studies for Mexican Agriculture, R. D. Norton and L. Solis M. (eds). Baltimore: Johns Hopkins University Press: 225-49.
Heady, E. O. 1952. Economics of Agricultural Production and Resource Use. Englewood Cliffs: Prentice Hall.
Heady, E. O., and D. R. Kaldor. 1954. "Expectations and Errors in Forecasting Agricultural Prices." Journal of Political Economy 62: 34-47.
Just, R. E. 1974. Econometric Analysis of Production Decisions with Government Intervention: The Case of the California Field Crops. Berkeley: University of California, Giannini Foundation Monograph No. 33.
_____. 1975. "Risk Response Models and Their Use in Agricultural Policy Evaluation." American Journal of Agricultural Economics 57: 836-43.
_____. 1976. "Estimation of a Risk Response Model With Some Degree of Flexibility." Southern Economic Journal 42: 633-43.
_____. 1977. "Estimation of an Adaptive Expectations Model." International Economic Review 18: 629-44.
_____. 1978. The Welfare Economics of Agricultural Risk. Berkeley: University of California, Giannini Foundation, mimeograph.
Just, R. E., E. Lutz, A. Schmitz, and S. J. Turnovsky. 1978. "The Distribution of Welfare Gains from Price Stabilization." Journal of International Economics 8: 551-63.
Just, R. E., and A. Schmitz. 1979. Nonoptimality of Price Bands in Stabilization Policy. Berkeley: University of California, Giannini Foundation, mimeograph.
Kaldor, D. R., and E. O. Heady. 1954. An Exploratory Study of Expectations, Uncertainty and Farm Plans in Southern Iowa Agriculture. Ames: Iowa State University, Agricultural Experiment Station Research Bulletin No. 408.
Kelley, A. C., J. G. Williamson, and R. J. Cheetham. 1972. Dualistic Economic Development: Theory and History. Chicago: University of Chicago Press.

Little, I. M. D., and J. A. Mirrlees. 1974. Project Appraisal and Planning for Developing Countries. London: Heinemann.

Lutz, E. 1978. "The Welfare Gains from Price Stabilization Under Risk Response." Schweig Zeitschrift fur Volkawirtschaft, und Statutk 2: 115-30.

MacKinnon, J. G. 1983. "Model Specification Tests Against Non-nested Alternatives." Econometric Review 2: 85-110.

Magnûsson, G. 1969. Production Under Risk, A Theoretical Study. Upsala: ACTA, Universitatis Upsaliensis.

Massell, B. F. 1969. "Price Stabilization and Welfare." Quarterly Journal of Economics 83: 284-98.

Moscardi, E., and A. de Janvry. 1978. "Attitudes Toward Risk Among Peasants: An Econometric Approach." American Journal of Agricultural Economics 59: 710-16.

Muth, J. F. 1961. "Rational Expectations and the Theory of Price Movements." Econometrica 29: 315-35.

Nerlove, M. 1956. "Estimates of the Elasticities of Supply of Selected Agricultural Commodities." Journal of Farm Economics 38: 496-509.

_____. 1958. The Dynamics of Supply: Estimation of Farmers' Response to Price. Baltimore: Johns Hopkins University Press.

Newbery, D. M. G. 1976. Price Stabilization with Risky Production. Stanford: Institute for Mathematical Studies in the Social Sciences, Stanford University Economic Series No. 69.

Newbery, D. M. G., and J. E. Stiglitz. 1979. "The Theory of Commodity Price Stabilization Rules: Welfare Impacts and Supply Responses." Economic Journal 89: 799-817.

_____. 1981. The Theory of Commodity Price Stabilization. New York: Clarendon Press; Oxford University Press.

_____. 1982. "The Choice of Techniques and the Optimality of Market Equilibrium With Rational Expectations." Journal of Political Economy 90: 223-46.

Norton, R. D., and G. W. Schiefer. 1980. "Agricultural Sector Programming Models: A Review." European Review of Agricultural Economics 7: 229-65.

Oi. W. Y. 1961. "The Desirability of Price Instability Under Perfect Competition." Econometrica 29: 58-64.

Peck, A. E. 1976. "Futures Markets, Supply Response, and Price Stability." Quarterly Journal of Economics 90: 407-23.

Pereira, B. de B. 1977. "A Note on the Consistency and on the Finite Sample Comparisons of Some Tests of Separate Families of Hypotheses." Biometrika 64: 109-13.

Pesaran, M. H. 1974. "On the General Problems of Model Selection." Review of Economic Studies 42: 153-71.

Pesaran, M. H., and A. S. Deaton. 1978. "Testing Non-Nested Nonlinear Regression Models." Econometrica 46: 677-94.

Pomerada, C., (ed). 1980. Estudio de Desarrollo Agricola de Centroamerica, Vol. 1. Guatemala: ECID., Secretaria Permanente del Tratado General de Integracion Economica Centroamericana.

Pope, R. D., and R. E. Just. 1977. "On the Competitive Price Under Production Uncertainty." Australian Journal of Agricultural Economics 21: 111-18.

Pouliquen, L. Y. 1970. Risk Analysis in Project Appraisal. Washington, D.C.: World Bank Staff Occasional Papers No. 11.

Powell, A. A., and F. H. Gruen. 1966a. "Problems in Aggregate Agricultural Supply Analysis: I, The Construction of Time Series for Analysis." Review of Marketing and Agricultural Economics 34: 112-35.

_____. 1966b. "Problems in Aggregate Agricultural Supply Analysis: II, Preliminary Results for Cereals and Wool." Review of Marketing and Agricultural Economics 34: 186-201.

_____. 1968. "The Constant Elasticity of Transformation Production Frontier and Linear Supply System." International Economic Review 9: 315-28.

Quiggin, J. C. and J. R. Anderson. 1979. "Stabilization and Risk Reduction in Australian Agriculture." Australian Journal of Agricultural Economics 23: 191-206.

Reutlinger, S. 1964. Evaluation of Some Uncertainty Hypotheses for Predicting Supply. Raleigh: North Carolina Agricultural Experiment Station, Technical Bulletin No. 160.

_____. 1970. Techniques for Project Appraisal Under Uncertainty. Washington, D.C.: World Bank Staff Occasional Papers No. 10.

Roumasset, J. A., J. M. Boussard, and I. Singh, (eds). 1979. Risk, Uncertainty and Agricultural Development. Laguna, Philippines: Southeast Asian Regional Center for Graduate Study and Research in Agriculture, and the Agricultural Development Council.

Ryan, T. J. 1977. "Supply Response to Risk: The Case of U. S. Pinto Beans." Western Journal of Agricultural Economics 2: 35-43.

Sanderson, B. A., J. J. Quilkey, and J. W. Freebairn. 1980. "Supply Response of Australian Wheat Growers." Australian Journal of Agricultural Economics 24: 129-40.

Sandmo, A. 1971. "On the Theory of the Competitive Firm Under Price Uncertainty." American Economic Review 61: 65-73.

Samuelson, P. A. 1952. "Spatial Price Equilibrium and Linear Programming." American Economic Review 42: 283-303.

Scandizzo, P. L., P. B. R. Hazell and J. R. Anderson. 1983. "Producers' Price Expectations and the Size of the Welfare Gains from Price Stabilization." Review of Marketing and Agricultural Economics 51: 93-107.

Schaller, W. N. 1969. "Discussion: The Supply Function in Agriculture Revisited." American Journal of Agricultural Economics 51: 367-69.

Schmalensee, R. 1976. "An Experimental Study of Expectation Formations." Econometrica 44: 117-41.

Schultz, T. W., and O. H. Brownlee. 1942. "Two Trials to Determine Expectation Models Applicable to Agriculture." Quarterly Journal of Economics 56: 487-96.

Seagraves, J. A. 1970. Agricultural Production and Supply Response Research: A Review for Project Planners in Developing Countries. Washington, D.C.: World Bank Economics Department Working Paper No. 57.

Sengupta, J. K., and A. Sen. 1969. "Econometric Supply Functions for Rice and Jute." Arthaniti 12: 1-40.

Smyth, D. J. 1973. "Effect of Public Price Forecasts on Market Price Variation: A Stochastic Cobweb Example." American Journal of Agricultural Economics 55: 83-88.

Takayama, T., and G. G. Judge. 1964. "Spatial Equilibrium and Quadratic Programming." Journal of Farm Economics 46: 67-93.

_____. 1971. Spatial and Temporal Price and Allocation Models. Amsterdam: North-Holland.

Thomas, W., L. Blakeslee, L. Rogers, and N. Whittlesey. 1972. "Separable Programming for Considering Risk in Farm Planning." American Journal of Agricultural Economics 54: 260-66.

Thomson, K. J., and P. B. R. Hazell. 1972. "Reliability of Using the Mean Absolute Deviation to Derive Efficient E, V Farm Plans." American Journal of Agricultural Economics 54: 503-06.

Tomek, W. G., and K. L. Robinson. 1977. "Agricultural Price Analysis and Outlook." In A Survey of Agricultural Economics Literature Volume 1, L. R. Martin (ed). Minneapolis: University of Minnesota Press: 329-409.

Traill, B. 1978. "Risk Variables in Econometric Supply Response Models." Journal of Agricultural Economics 24: 53-61.

Turnovsky, S. J. 1973. "Production Flexibility, Price Uncertainty and the Behavior of the Competitive Firm." International Economic Review 14: 395-413.

_____. 1974. "Price Expectations and the Welfare Gains from Price Stabilization." American Journal of Agricultural Economics 56: 706-16.

134

_____. 1976. "The Distribution of Welfare Gains from Price Stabilization: The Case of Multiplicative Disturbances." International Economic Review 17: 133-48.

_____. 1978. "The Distribution of Welfare Gains from Price Stabilization: A Survey of Some Theoretical Issues." In Stabilizing World Commodity Markets, F. G. Adams and S. A. Klein (eds). Lexington: Lexington Books: 119-48.

Waugh, F. V. 1944. "Does the Consumer Benefit from Price Instability?" Quarterly Journal of Economics 58: 602-14.

Willig, R. D. 1976. "Consumer's Surplus Without Apology." American Economic Review 66: 589-97.

Williams, D. B. 1951. "Price Expectations and Reactions to Uncertainty by Farmers in Illinois. Journal of Farm Economics 33: 20-39.

World Bank. 1978. World Development Report, 1978. Washington, D.C.: World Bank.

Wright, B. D. 1979. "The Effects of Total Production Stabilization: A Welfare Analysis Under Rational Behavior." Journal of Political Economy 87: 1011-33.

Zusman, P. 1969. "The Stability of Interregional Competition and the Programming Approach to the Analysis of Spatial Trade Equilibria." Metroeconomica 11: 45-57.

Author Index

135

Subject Index